How to Guard an Art Gallery

How to Guard an Art Gallery

And Other Discrete Mathematical Adventures

T. S. Michael

The Johns Hopkins University Press
Baltimore

Printed in the United States of America on acid-free paper
9 8 7 6 5 4 3 2 1

The Johns Hopkins University Press
2715 North Charles Street
Baltimore, Maryland 21218-4363
www.press.jhu.edu

Library of Congress Cataloging-in-Publication Data

Michael, T. S., 1960–
How to guard an art gallery and other discrete mathematical adventures / T. S. Michael.
p. cm.
Includes bibliographical references and index.
ISBN-13: 978-0-8018-9298-1 (hardcover : alk. paper)
ISBN-10: 0-8018-9298-8 (hardcover : alk. paper)
ISBN-13: 978-0-8018-9299-8 (pbk. : alk. paper)
ISBN-10: 0-8018-9299-6 (pbk. : alk. paper)
1. Combinatorial analysis. 2. Algorithms. 3. Computer science—Mathematics. I. Title.
QA164.M53 2009
511'.6—dc22 2009000435

A catalog record for this book is available from the British Library.

Special discounts are available for bulk purchases of this book. For more information, please contact Special Sales at 410-516-6936 or specialsales@press.jhu.edu.

The Johns Hopkins University Press uses environmentally friendly book materials, including recycled text paper that is composed of at least 30 percent post-consumer waste, whenever possible. All of our book papers are acid-free, and our jackets and covers are printed on paper with recycled content.

Contents

Preface

The adventures in this book are launched by easily understood questions from the realm of *discrete mathematics,* a wide-ranging subject that studies fundamental properties of the counting numbers 1, 2, 3, ... and arrangements of finite sets

The book grew from talks for mathematically inclined secondary school students and college students interested in problem solving. The aim is high, but the prerequisites are modest—mostly elementary algebra and geometry. Occasionally, a perspective gained from more advanced subjects is mentioned. A sampling of questions conveys the spirit and scope of the topics.

- **The art gallery problem.** What is the minimum number of stationary guards (or security cameras) needed to protect a given art gallery?
- **The pizza-cutter's problem.** What is the maximum number of pizza pieces we can make with four straight cuts through a circular pizza? What about n cuts?
- **The computer line drawing problem.** Which pixels should a computer select to represent a given straight line on a monitor?
- **A quadratic residue question.** Is there an integer whose square is 257 more than a multiple of 641? In the jargon of number theory, is 257 a quadratic residue modulo 641?

Our interest extends beyond answers to individual questions, no matter how accessible and enticing. The questions are gateways to deeper mathematical material that can be discussed without a lot of background. For instance, the following puzzle (taken from a memorable scene in the movie

Die Hard: With a Vengeance) leads to a discussion of a famous result of Fermat in number theory.

- **The Bruce Willis problem.** We are at a fountain with two unmarked jugs with capacities 3 and 5 gallons. How can we measure exactly 4 gallons of water?

Our goal is to impart a genuine feel for discovery and mature mathematical thinking by attacking problems from several points of view and in various degrees of generality. We also reveal hidden connections between seemingly unrelated topics. For instance, we will discover a relationship between computer line drawing and quadratic residues. You will also likely be surprised to learn that the following two questions are related.

- **An area question.** What is the area of the oddly shaped orchard shown in the figure if the rows and columns of trees are 1 unit apart?

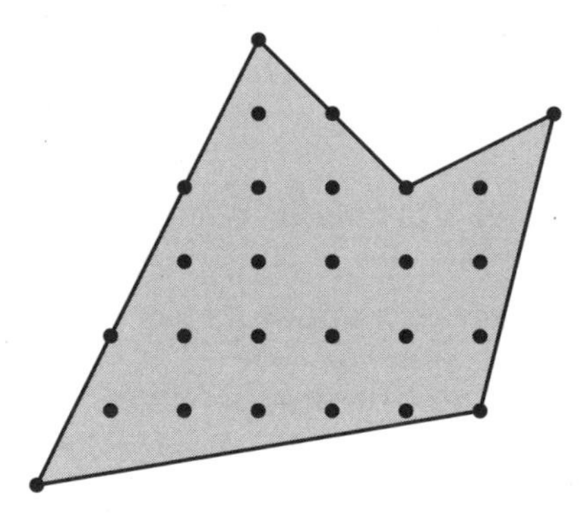

- **A dollar-changing question.** How many ways are there to make change for a dollar with quarters, dimes, and nickels?

Readers inspired to chart their own mathematical adventures can explore the problems at the end of each chapter. The more challenging problems include hints or are broken into smaller steps. The lightly annotated references are a starting point for further reading.

Among the many people who helped and encouraged me as I wrote this book, several deserve special thanks. Amy Myers, Courtney Moen, and Sommer Gentry gave me valuable feedback on all aspects of early drafts and pointed out ways to improve each chapter. I also greatly appreciate the patience and guidance of my editor, Trevor Lipscombe.

Finally, I dedicate this book to Tom Apostol, who set me on the path to mathematical maturity 30 years ago.

How to Guard an Art Gallery

1

How to Count Pizza Pieces

You better cut the pizza in four pieces
because I'm not hungry enough to eat six.
YOGI BERRA

1.1 The Pizza-Cutter's Problem

Three cuts through the center of a circular pizza make six identical pieces of pizza. Displacing one cut slightly gives seven pieces of various sizes and shapes, as in Figure 1.1. A

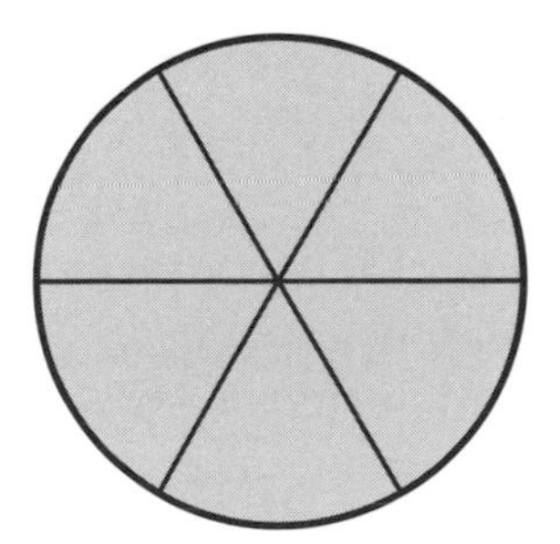

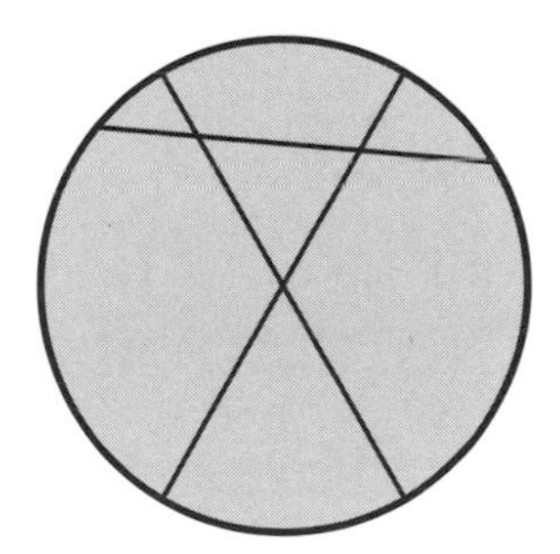

Figure 1.1: Three cuts through a pizza

little experimentation should convince you that seven pieces is the most you can make with three cuts. But how many pizza pieces can you make with more cuts?

The pizza-cutter's problem. What is the largest number of pieces of pizza we can make with n straight cuts through a circular pizza?

The goal is to maximize the number of pieces regardless of size and shape. In this chapter, we serve up several solutions to the pizza-cutter's problem. Our solutions introduce a host of ideas from discrete mathematics. The first solution is based on a recurrence, which gives the number of pizza pieces with n cuts in terms of the number with $n-1$ cuts. The second solution uses a table of differences to guide us to the general form of the answer. Later approaches relate the number of pizza pieces to the number of points where two cuts cross. One solution consists of a single diagram!

Setting the Table: Small Values

Let $P(n)$ denote the maximum number of pieces formed by n straight cuts through a circular pizza. We want to find a formula for $P(n)$. The first order of business is to gather some data by cutting pizzas or drawing lines through circles. It is easy to see that

$$P(0) = 1, \quad P(1) = 2, \quad \text{and} \quad P(2) = 4.$$

Figures 1.1 and 1.2 help us see that

$$P(3) = 7, \quad P(4) = 11, \quad \text{and} \quad P(5) = 16.$$

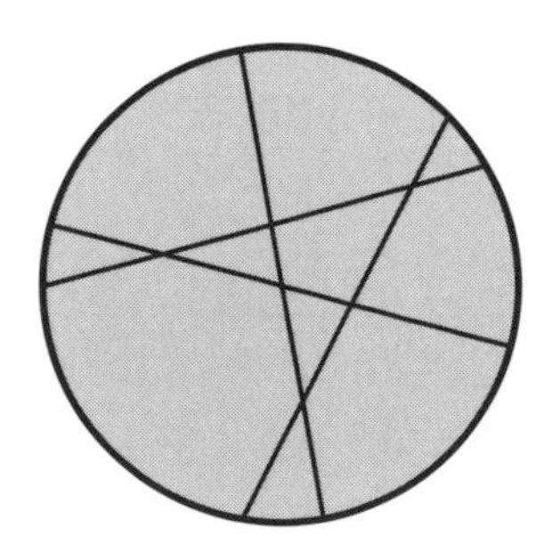
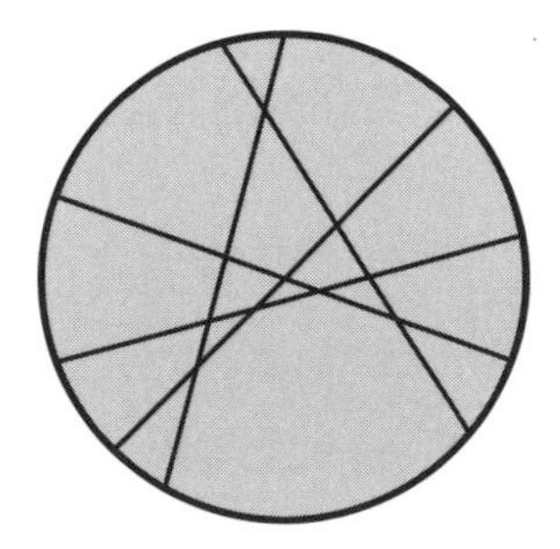

Figure 1.2: Four and five cuts through a pizza

Table 1.1: The maximum number of pizza pieces with n cuts

n	0	1	2	3	4	5	6	7
$P(n)$	1	2	4	7	11	16	22	29

Optimal configurations become more difficult to draw as the number of cuts increases. It might take you several attempts to make 29 pieces with seven cuts. Table 1.1 lists the values of $P(n)$ for $n = 0, 1, \ldots, 7$. As you construct the optimal configurations, the following principle becomes clear.

Maximizing principle. The number of pizza pieces is maximized when every cut crosses every other cut, but no three cuts cross at the same point.

Jakob Steiner's Plane Pizza

The pizza-cutter's problem was first studied and solved in an equivalent form by the prominent Swiss geometer Jakob Steiner (1796–1863). Steiner's pizza was infinitely large—occupying the whole plane—and his cuts were infinitely long straight lines.

Steiner's plane-cutting problem. What is the maximum number of regions formed by n lines in the plane?

Every line intersects every other line in an optimal configuration. Note that some of the regions will be infinitely large.

Figure 1.3 shows why Steiner's plane-cutting problem is equivalent to the pizza-cutter's problem. Start with a configuration of n cuts that intersect pairwise inside a circular

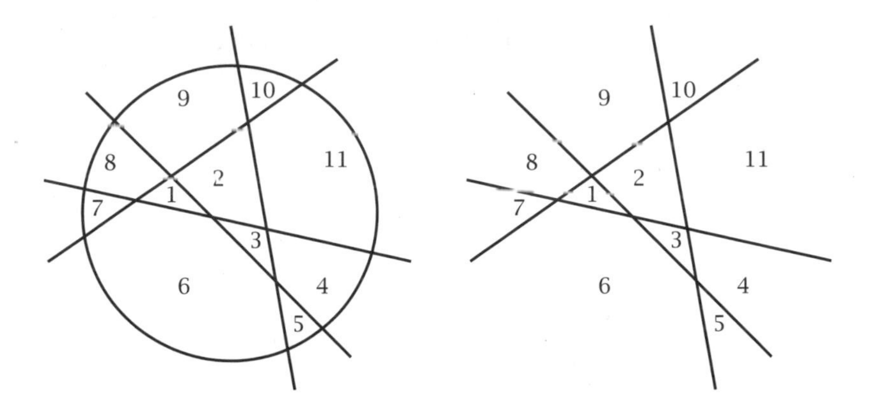

Figure 1.3: Pizza-cutting and Steiner's problem

pizza to form $P(n)$ pieces. Then extend the cuts indefinitely in both directions and delete the circular boundary of the pizza. The resulting configuration of n straight lines partitions the entire plane into $P(n)$ regions, as shown in the figure. The "crusty" pizza pieces (those touching the circular boundary) become regions that extend infinitely far, but no regions are created or destroyed. The process works in reverse, too. If n lines form a certain number of regions in the plane, then adding a suitably large boundary circle gives the same number of finite pizza pieces.

We find it more convenient to work in the pizza-cutting scenario than in Steiner's purely geometric setting.

1.2 A Recurring Theme

Perhaps you noticed a pattern among the numbers

$$1, 2, 4, 7, 11, 16, 22, 29, \ldots$$

in Table 1.1. The differences between successive values are

$$1, 2, 3, 4, 5, 6, 7, \ldots.$$

This pattern can be expressed as an equation giving $P(n)$ in terms of $P(n-1)$, that is, as a *recurrence*.

Pizza-cutter's recurrence. The maximum number of pizza pieces formed by n cuts satisfies

$$P(n) = P(n-1) + n \qquad \text{for } n = 1, 2, 3, \ldots.$$

The recurrence lets us extend our table of values for $P(n)$ without actually cutting pizzas or drawing pictures. For instance, with eight cuts the maximum number of pizza pieces is

$$P(8) = P(7) + 8 = 29 + 8 = 37.$$

Such computations assume our recurrence is valid in general. Even though it holds for the first few values of n, it could fail for larger n. Discrete mathematics is rife with patterns that fall apart after several terms. For instance, Problem 7 at the end of this chapter deals with a geometric problem giving rise to a sequence that starts 1, 2, 4, 8, 16. But the next term is 31, not 32.

New Pizza Pieces from Old

Here is why the pizza-cutter's recurrence is true. Start with $n-1$ cuts that form $P(n-1)$ pieces of pizza. Slowly make the n-th cut (the dotted line in Figure 1.4) to form $P(n)$ pieces. When the new cut meets one of the $n-1$ previous ones, a pizza piece is cut in two. Also, a piece is cut in two when the new cut finishes on the opposite side of the pizza. So the total number of pieces of pizza increases by n when we pass from $n-1$ cuts to n cuts, which is exactly what the recurrence says.

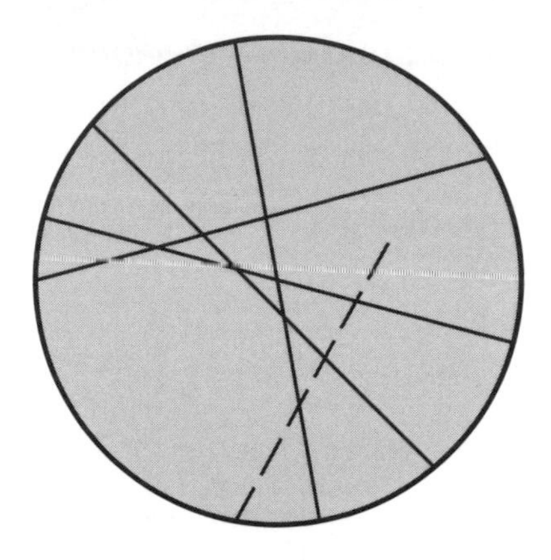

Figure 1.4: The n-th cut produces n more pieces of pizza

The Pizza-Cutter's Formula

We can replace n by $n-1$ in our recurrence to write $P(n-1)$ in terms of $P(n-2)$:

$$P(n-1) = P(n-2) + (n-1).$$

In general, we can write $P(k)$ in terms of $P(k-1)$ for $k = n$ all the way down to $k = 1$:

$$\begin{aligned}
P(n) &= P(n-1) + n \\
P(n-1) &= P(n-2) + (n-1) \\
P(n-2) &= P(n-3) + (n-2) \\
&\vdots \\
P(3) &= P(2) + 3 \\
P(2) &= P(1) + 2 \\
P(1) &= P(0) + 1.
\end{aligned}$$

Add all n equations and cancel the common terms from both sides to obtain

$$P(n) = P(0) + 1 + 2 + \cdots + (n-2) + (n-1) + n.$$

Now we recall a well-known formula for the sum of the first n positive integers:

$$1 + 2 + 3 + \cdots + n = \frac{n(n+1)}{2}. \tag{1.1}$$

Therefore,

$$P(n) = P(0) + \frac{n(n+1)}{2} = 1 + \frac{n(n+1)}{2} = \frac{n^2 + n + 2}{2}.$$

We have solved the pizza-cutter's problem by essentially the same method Steiner used.

Pizza-cutter's formula. The maximum number of pieces we can make with n straight cuts through a circular pizza is

$$P(n) = \frac{n^2 + n + 2}{2}.$$

If you dislike recurrences, or if formula (1.1) is not familiar to you, then our first solution to the pizza-cutter's problem might be unsatisfying. Perhaps one of the other solutions in this chapter is more to your taste. In any case, the alternative solutions provide more insight and suggest different generalizations.

1.3 Make a Difference

The *calculus of finite differences* is a powerful tool that uses tables of numbers to explore patterns in sequences of numbers. Table 1.2 is the *difference table* for the pizza-cutting sequence $P(0), P(1), P(2), \ldots$. The top row is the sequence itself, and each successive row is obtained by subtracting consecutive terms from the row above. The pizza-cutter's recurrence guarantees that the row of first differences is correct, and it follows that all rows in the table are correct.

Table 1.2: The difference table for $P(n)$

sequence	1		2		4		7		11		16		22	·
first differences		1		2		3		4		5		6		·
second differences			1		1		1		1		1		·	·
third differences				0		0		0		0		·		·

The row of third differences in Table 1.2 consists of 0's, and the theory tells us that $P(n)$ must be a quadratic function[1] of n, say,

$$P(n) = An^2 + Bn + C, \tag{1.2}$$

for some constants A, B, and C. Our known values of $P(0)$, $P(1)$, and $P(2)$ lead to the system

$$\left\{ \begin{array}{rcccccccc} 1 & = & P(0) & = & & & & & C \\ 2 & = & P(1) & = & A & + & B & + & C \\ 4 & = & P(2) & = & 4A & + & 2B & + & C \end{array} \right\},$$

which we solve to find

$$A = 1/2, \quad B = 1/2, \quad \text{and} \quad C = 1.$$

Now our quadratic function in (1.2) becomes our pizza-cutter's formula

$$P(n) = \frac{1}{2}n^2 + \frac{1}{2}n + 1 = \frac{n^2 + n + 2}{2}.$$

If you want to try your hand with difference tables, use Table 1.3 to confirm formula (1.1) for the sum $S(n) = 0 + 1 + 2 + \cdots + n$.

[1]If the row of k-th differences for a sequence $f(0)$, $f(1)$, $f(2)$, ... contains only 0's, then the calculus of finite differences tells us that $f(n)$ is a polynomial in the variable n of degree at most $k - 1$. There is an analogous result in differential calculus: If the k-th derivative of a function $f(x)$ is identically 0, then $f(x)$ must be a polynomial in x of degree at most $k - 1$.

Table 1.3: The difference table for $S(n) = 0 + 1 + 2 + \cdots + n$

sequence	0	1	3	6	10	15	21	·
first differences	1	2	3	4	5	6	·	
second differences	1	1	1	1	1	·	·	
third differences	0	0	0	0	·	·		

1.4 How Many Toppings?

The argument that produced our basic recurrence (see Figure 1.4) hints at a relationship between the number of pizza pieces and the number of intersection points formed by the cuts. This relationship is the key to another solution to the pizza-cutter's problem.

Figure 1.5 illustrates the main idea. We have temporarily dropped the requirement that every pair of cuts intersect. But we continue to assume that no three cuts intersect at

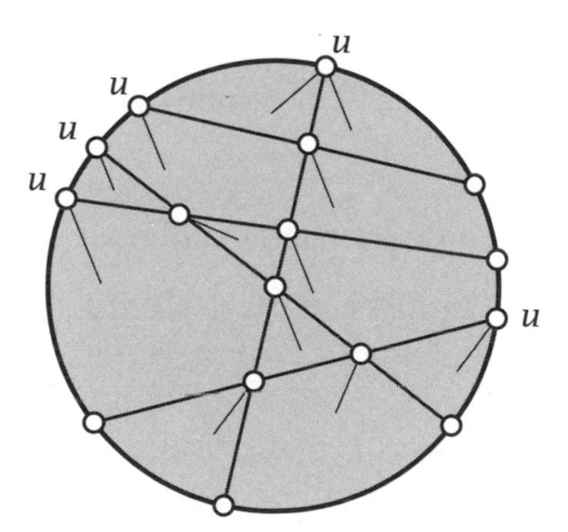

Figure 1.5: The top vertex of each pizza piece is either an interior vertex or an upper boundary vertex (u)

the same point and that no two cuts intersect at a boundary point. An *interior vertex* occurs where two cuts intersect. Also, each of the n cuts determines an *upper boundary vertex*, as labeled in the figure. We rotate the pizza if necessary so that no cut is horizontal, guaranteeing that each pizza piece has a unique top vertex.

The tails suspended from some vertices in the figure match each piece of pizza to its top vertex. Every upper boundary vertex is matched with one pizza piece, except for the highest boundary vertex, which is matched with two pieces. Also, every interior vertex is matched with one piece. Our matches give a formula to count pizza pieces:

$$\begin{array}{c}\text{number of}\\\text{pizza pieces}\end{array} = 1 + \begin{array}{c}\text{number}\\\text{of cuts}\end{array} + \begin{array}{c}\text{number of}\\\text{interior vertices}\end{array}.$$

There are five cuts and six interior vertices in Figure 1.5, and so the formula tells us that the number of pizza pieces is $1 + 5 + 6 = 12$, which is easy to verify.

To maximize the number of pizza pieces, each pair of cuts must cross at a unique interior vertex. In this case, we have

$$\begin{array}{c}\text{maximum}\\\text{number of}\\\text{pizza pieces}\end{array} = 1 + \begin{array}{c}\text{number}\\\text{of cuts}\end{array} + \begin{array}{c}\text{number of}\\\text{pairs of cuts}\end{array}.$$

This brings us to a new way to write the pizza-cutting formula using the *choose function* ("n choose k"), defined by

$$\binom{n}{k} = \begin{array}{l}\text{the number of ways to choose a subset}\\\text{of } k \text{ elements from a set of } n \text{ elements.}\end{array}$$

Pizza-cutter's formula (new version). The maximum number of pieces we can make with n straight cuts through a circular pizza is

$$P(n) = \binom{n}{0} + \binom{n}{1} + \binom{n}{2}.$$

Let us see why the new version is equivalent to the original pizza-cutter's formula. From a set of n cuts, there is 1 way to choose 0 cuts (namely, do nothing) and n ways to choose 1 cut. Also, in choosing a pair of cuts, there are n choices for the first cut and $n-1$ choices for the second cut, giving $n(n-1)/2$ pairs of cuts altogether. We divide by 2 because the two cuts can be chosen in either order. Therefore,

$$\binom{n}{0} + \binom{n}{1} + \binom{n}{2} = 1 + n + \frac{n(n-1)}{2} = \frac{n^2+n+2}{2},$$

which is our original pizza-cutter's expression.

From Pizza to Grapefruit and Beyond

The new version of the pizza-cutter's formula reveals its place within a hierarchy of similar formulas in other dimensions. Here is the three-dimensional case, which was also discovered by Steiner.

Grapefruit cutter's formula. The maximum number of pieces of fruit we can make with n straight cuts through a spherical grapefruit is

$$\binom{n}{0} + \binom{n}{1} + \binom{n}{2} + \binom{n}{3}.$$

Each cut is a plane that passes completely through the grapefruit. One way to establish the formula relies on a difference table. See Problem 17.

Even if you are unable to visualize a solid d-dimensional sphere being sliced by $(d-1)$-dimensional cuts, you will not be surprised to learn that the number of pieces formed by n cuts is

$$\binom{n}{0} + \binom{n}{1} + \binom{n}{2} + \cdots + \binom{n}{d-1} + \binom{n}{d}.$$

Challenge 1. Explain why the formula makes sense when $d = 1$. A one-dimensional "sphere" is just a line segment, and each cut is a point on the segment.

We remark that the choose function can be computed with the formula

$$\binom{n}{k} = \frac{n!}{k!\,(n-k)!} \qquad \text{for } k = 0, 1, \ldots, n,$$

where $m!$ ("m factorial") is defined by

$$m! = 1 \times 2 \times 3 \times \cdots \times (m-1) \times m \qquad \text{for } m = 1, 2, 3, \ldots$$

and $0! = 1$.

1.5 Proof without Words

We now return to our two-dimensional pizza. Figure 1.6 encapsulates in a single picture our earlier vertex-counting argument. The figure by itself constitutes a visual proof—a *proof without words*—of the pizza-cutter's formula. In case this type of demonstration is unfamiliar to you, we include some words of explanation.

The figure illustrates the case $n = 6$, but it will be apparent that the idea works generally. Start with n lines $L_1, L_2, \ldots, L_n$ that cut the pizza into $P(n)$ pieces. The label for each piece describes its highest point. The highest point of a piece labeled with the single number i is where L_i meets the circular edge of the pizza. There are $\binom{n}{1}$ such singletons, except for the two uppermost pieces, which have the same label. The term $\binom{n}{0}$ accounts for the extra label. Also, the intersection of lines L_i and L_j is the highest point for the piece labeled with the pair i, j. There are $\binom{n}{2}$ such pairs. The figure accounts for all terms in the formula $P(n) = \binom{n}{0} + \binom{n}{1} + \binom{n}{2}$.

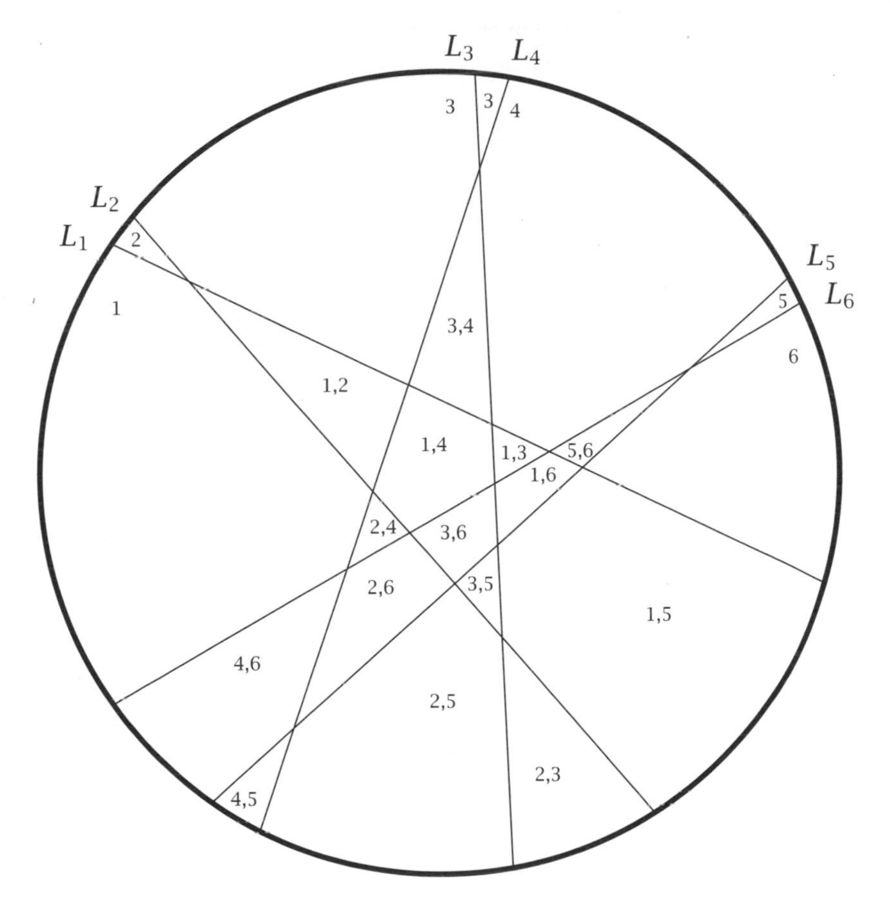

Figure 1.6: A proof without words: $P(n) = \binom{n}{0} + \binom{n}{1} + \binom{n}{2}$

Challenge 2. Explain how Figure 1.7 gives a proof without words for the formula we used in our discussion of recurrence relations:

$$1 + 2 + 3 + \cdots + n = \frac{n(n+1)}{2}.$$

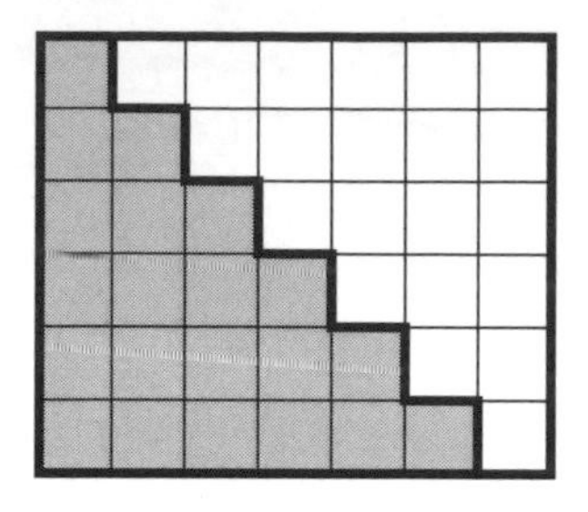

Figure 1.7: A proof without words: $1+2+3+\cdots+n = n(n+1)/2$

1.6 Count 'em and Sweep

Figure 1.8 shows a pizza that has been cut into pieces in a general manner. Cuts are no longer required to pass all the way through the pizza. A *vertex* occurs wherever two or more cuts meet or where a cut meets the circular boundary of the pizza. An *edge* is the segment or arc joining two consecutive vertices on a cut or two consecutive vertices on the boundary. We let v, e, and p denote the number of vertices, edges, and pizza pieces formed by a given arrangement of cuts. For instance, $v = 15$, $e = 30$, and $p = 16$ in Figure 1.8.

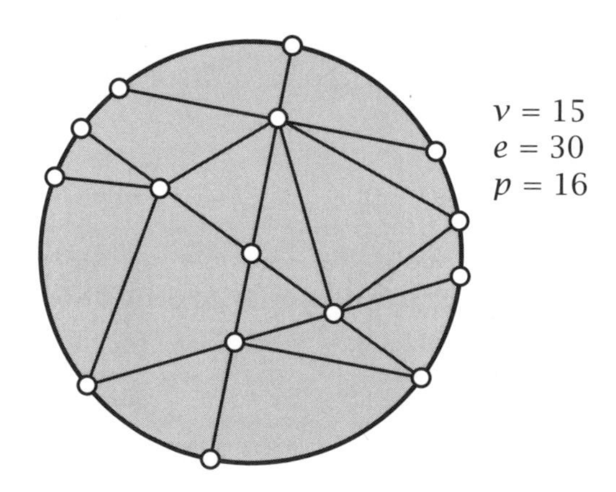

Figure 1.8: Vertices, edges, and pizza pieces

Our work with topmost vertices foreshadows a fundamental relationship among v, e, and p. This relationship can be discovered and explained using a horizontal *sweep-line,* sweeping upward from the bottom to the top of the pizza, as indicated in Figure 1.9. We let $\overline{v}$, $\overline{e}$, and $\overline{p}$ be counters that keep track of the number of vertices, edges, and pieces below or on the sweep-line. (The bar above each parameter is a visual reminder of the horizontal sweep-line.)

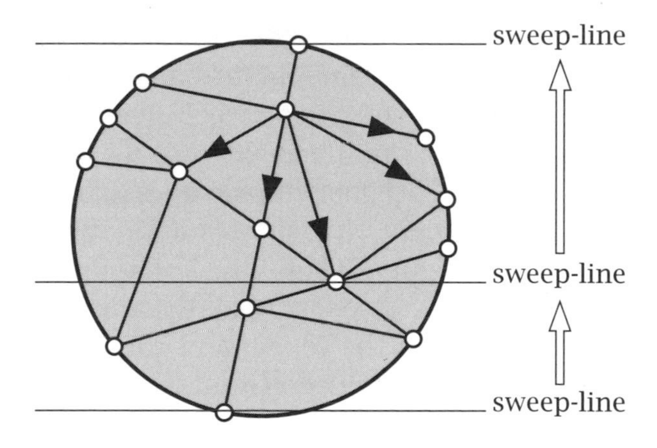

Figure 1.9: A horizontal sweep-line

The sweep-line passes through the lowest vertex at the outset, and so the counters start with the values

$$\overline{v} = 1, \quad \overline{e} = 0, \quad \text{and} \quad \overline{p} = 0.$$

Note that $\overline{v} - \overline{e} + \overline{p} = 1$. The counters increase each time the sweep-line reaches a new vertex. If the vertex is joined by E edges to lower vertices, as indicated by arrows in Figure 1.9, then $\overline{v}$ increases by 1, $\overline{e}$ increases by E, and $\overline{p}$ increases by $E-1$. It follows that the expression $\overline{v}-\overline{e}+\overline{p}$ does not change during the sweeping process and is thus always equal to 1. All vertices, edges, and pieces are accounted for when the sweep-line passes through the top vertex.[2] We have shown that the counts of vertices, edges, and pizza pieces satisfy $v - e + p = 1$. For our purposes, this relation is more conveniently written

$$p = e - v + 1. \tag{1.3}$$

For instance, the pizza in Figure 1.9 satisfies

$$p = 16 = 30 - 15 + 1 = v - e + 1.$$

There is one potential glitch in the preceding argument. Our horizontal sweep-line might encounter two vertices simultaneously. This situation can always be circumvented by rotating the pizza slightly at the outset.

To solve our original pizza-cutter's problem with formula (1.3), we count the vertices and edges in an optimal configuration of cuts. There are $2n$ boundary vertices and $n(n-1)/2$ interior vertices. Also, there are $2n$ curved edges on the boundary of the pizza, and each of the n cuts contains n straight edges. Therefore,

$$v = 2n + \frac{n(n-1)}{2} \qquad \text{and} \qquad e = 2n + n^2.$$

We use these expressions for v and e in (1.3) and find that the maximum number of pizza pieces with n cuts is the familiar expression $p = (n^2 + n + 2)/2$.

1.7 Euler's Formula for Plane Graphs

The preceding solution to the pizza-cutter's problem has brought us to the doorstep of an important formula discovered by the great Swiss mathematician Leonhard Euler (1707–1783).

---

[2]We regard the curved edges and pieces meeting the top vertex as falling below the sweep-line, even though they might bulge slightly above.

A *plane graph* consists of a finite set of *vertices* and a list of *edges,* which are pairs of distinct vertices. Each vertex is represented by a point in the plane, and each edge is represented by a segment or curve joining two vertices. The edges cannot cross or touch one another (except at common vertices), and so the plane is partitioned into regions called *faces,* one of which is infinite.

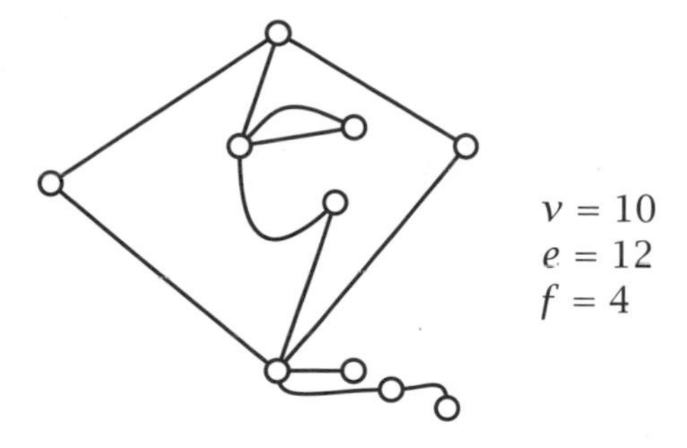

Figure 1.10: A connected plane graph with 10 vertices, 12 edges, and 4 faces

Plane graphs are used as models for transportation systems, chemical molecules, and electric circuits. A plane graph is *connected* provided one can travel from one vertex to any other vertex along edges. For instance, the plane graph in Figure 1.10 is connected, but deletion of any of the three lowest edges produces a disconnected plane graph. Let v, e, and f denote the number of vertices, edges, and faces of a plane graph. For the plane graph in Figure 1.10, we have $v = 10$, $e = 12$, and $f = 4$.

Euler's formula for plane graphs. If a connected plane graph has v vertices, e edges, and f faces, then

$$v - e + f = 2.$$

Because pizza diagrams of the type in Figure 1.9 define plane graphs, Euler's formula includes our earlier formula

$p = e - v + 1$ as a special case. Note that $f = p + 1$ since f counts the infinite region, too.

We remark that our earlier sweep-line argument does not quite prove Euler's formula in general. The presence of curved edges in plane graphs causes a few minor technical difficulties that a rigorous proof would treat carefully.

Going in Circles

We now apply Euler's formula to count the maximum number of regions formed by n circles in the plane, another situation studied by Steiner. The maximum is achieved when every pair of circles meets in two points, but no three circles pass through a common point. Figure 1.11 shows four circles that partition the plane into 14 regions.

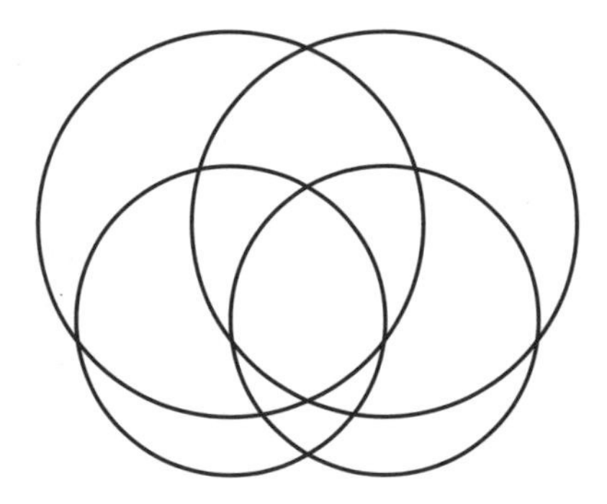

Figure 1.11: Four circles form 14 regions

If $n \geq 2$, the circles define a plane graph with the points of intersection serving as vertices and circular arcs as edges. It is not difficult to show that there are $v = n(n-1)$ vertices and $e = 2n(n-1)$ edges. By Euler's formula the number of regions is

$$f = e - v + 2 = n^2 - n + 2.$$

The formula also works for $n = 1$, and we have established the following result.

Theorem. The maximum number of regions formed by n circles in the plane is $n^2 - n + 2$ for $n = 1, 2, \ldots$.

Polyhedra

Euler discovered his fundamental formula while investigating *polyhedra*—three-dimensional solids bounded by polygonal faces. The most familiar polyhedron is the cube, which satisfies $v - e + f = 2$ since there are $v = 8$ vertices, $e = 12$ edges, and $f = 6$ faces. Figure 1.12 and Table 1.4 give a few other familiar examples.

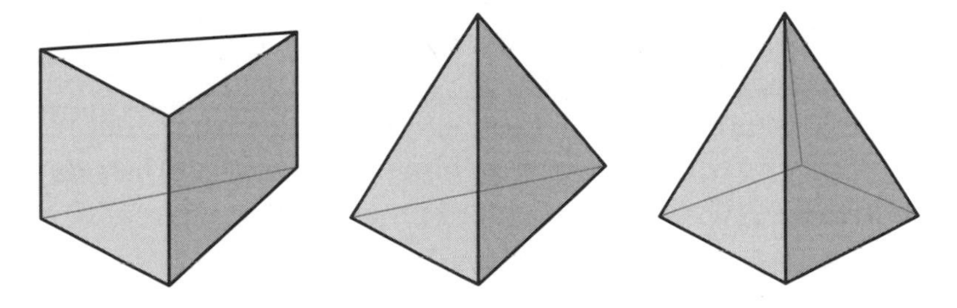

Figure 1.12: Triangular prism, triangular pyramid, square pyramid

Table 1.4: Polyhedra and Euler's formula

polyhedron	vertices	edges	faces	$v - e + f$
cube	8	12	6	2
triangular prism	6	9	5	2
triangular pyramid	4	6	4	2
square pyramid	5	8	5	2
pentagonal pyramid	6	10	6	2

1.8 You Can Look It Up

Our final solution to the pizza-cutter's problem is the least taxing. We simply look up our sequence in the On-Line Encyclopedia of Integer Sequences (OEIS), by entering the first few known values, say,

$$1, 2, 4, 7, 11, 16,$$

at the site

`research.att.com/~njas/sequences/` .

The encyclopedia responds with many candidate sequences, each of which starts with the six specified terms. The one we want is sequence A000124, described as the "maximal number of pieces formed when slicing a pancake with n cuts." Our pizza-cutter's formula is given explicitly. The OEIS also provides citations and other contexts in which our sequence has arisen, ranging from binary vectors, to permutations, to zonohedra (polyhedra with a special type of symmetry).

The encyclopedia is the brainchild of Neil J. A. Sloane of AT&T Laboratories. Sloane's diverse projects in pure and applied mathematics prompted him to compile and organize written lists of integer sequences several decades ago. The project evolved into an electronic database of more than 100,000 sequences from the familiar to the obscure. Users around the world submit and document hundreds of new sequences monthly. You may enjoy looking up your favorite sequences in the OEIS, but be forewarned. The encyclopedia elicits an unpredictable mixture of amazement, consternation, and delight. For instance, the terms

$$1, 2, 4, 8, 16$$

lead off a remarkable number of interesting sequences, including the one in Problem 7.

1.9 Pizza Envy

We end our pizza-cutting adventure by discussing a related geometric problem with an unexpected answer. Suppose we want to make n cuts and serve pizza to $P(n)$ selfish children. Configurations of cuts similar to those in Figure 1.2 are unacceptable because some pieces are much bigger than others.

Question. Can we arrange the n cuts so that the $P(n)$ pieces are the same size?

It turns out that the answer is *no.* Some children get bigger pieces than others if there are more than two cuts. In fact, the situation is much more dire. The pizza pieces can not even be approximately the same size when lots of cuts are made.

Pizza envy theorem. If n cuts through a circular pizza form $(n^2+n+2)/2$ pieces, then at least one of the pieces is more than $n/8$ times as big as the average-sized piece.

The theorem does not say much when there are fewer than eight cuts. However, with eight cuts some child gets more than his or her fair share of pizza, and with 16 cuts, some child gets more than twice as much as another.[3] Because the fraction $n/8$ becomes arbitrarily large, the pizza inequity becomes more extreme as more cuts are used.

Let us see why the theorem is true. Without loss of generality the radius of the pizza is 1 unit. The average area of a piece of pizza then satisfies

$$\text{average area} = \frac{\text{area of pizza}}{\text{number of pieces}} = \frac{\pi\,(1)^2}{(n^2+n+1)/2} < \frac{2\pi}{n^2}.$$

[3]One should order more pizza instead of trying to make $P(16) = 137$ pieces with 16 cuts.

However, in 2000 Dan Ismailescu used intricate trigonometric manipulations to establish a lower bound for the area of the biggest piece.

Ismailescu's theorem. If n cuts $(n \geq 1)$ of a unit circular pizza form $(n^2+n+2)/2$ pieces, then the area of the biggest piece is greater than $\pi/4n$.

Combine these results to see that the areas satisfy

$$\frac{\text{biggest area}}{\text{average area}} > \frac{\pi/4n}{2\pi/n^2} = \frac{n}{8},$$

which is what the pizza envy theorem asserts.

1.10 Notes and References

Modern treatments and extensions of Steiner's original work [16] include [1, 17, 18]. See Banks [2] for more details on the application of finite differences to pizza-cutting and related problems. Brualdi's text [4] includes a more rigorous treatment of finite differences. Chapter 1 of Cuoco's excellent book [7] covers difference tables from the basics to more advanced ideas and highlights several connections to other areas of mathematics. The book by Lovász, Pelikán, and Vesztergombi [10] includes several of our solutions to the pizza-cutter's problem. Wetzl's article [17] uses sweep-lines to generalize the pizza-cutter's formula to cases in which some cuts are parallel and three or more cuts intersect at a point. Also see [14].

Good introductions to Euler's formula occur in Chapter 1 of the popular book [3] by Beck, Bleicher, and Crowe and in most books on graph theory, e.g., [5]. Numerous proofs and references for Euler's formula are given at David Eppstein's webpage

`www.ics.uci.edu/~eppstein/junkyard/euler/` .

The book by Richeson [13] is devoted to Euler's formula and its important role in modern mathematics. The book by Cromwell [6] treats polyhedra in detail.

See Sloane's article [15] for more background on the On-Line Encyclopedia of Integer Sequences. Ismailescu's theorem appears in [8]. Nelson's compilations [11] and [12] contain dozens of proofs without words. Problem 9 is from the article by Larson [9].

1. Alexanderson, G. L., and Wetzel, J. E., Simple partitions of space, *Mathematics Magazine* **51** (1978), 220–225.
2. Banks, Robert B., *Slicing Pizzas, Racing Turtles, and Further Adventures in Applied Mathematics.* Princeton University Press, Princeton, New Jersey, 1999.
3. Beck, Anatole, Bleicher, Michael N., and Crowe, Donald W., *Excursions into Mathematics.* A. K. Peters, Natick, Massachusetts, 2000.
4. Brualdi, Richard A., *Introductory Combinatorics,* 4th ed. Pearson/Prentice-Hall, Upper Saddle River, New Jersey, 2004.
5. Chartrand, Gary, and Lesniak, Linda, *Graphs & Digraphs,* 4th ed. Chapman & Hall/CRC, Boca Raton, Florida, 2005.
6. Cromwell, Peter R., *Polyhedra,* New ed. Cambridge University Press, Cambridge, U.K., 1999.
7. Cuoco, Al, *Mathematical Connections.* Mathematical Association of America, Washington, DC, 2005.
8. Ismailescu, D., Slicing the pie, *Discrete and Computational Geometry* **30** (2003), 263–276.
9. Larson, L., A discrete look at $1 + 2 + \cdots + n$, *College Mathematics Journal* **16** (1985), 369–382.
10. Lovász, László, Pelikán, József, and Vesztergombi, Katalin, *Discrete Mathematics: Elementary and Beyond.* Springer, New York, 2003.
11. Nelson, Roger B., *Proofs without Words: Exercises in Visual Thinking.* Mathematical Association of America, Washington, DC, 1993.
12. Nelson, Roger B., *Proofs without Words II: More Exercises in Visual Thinking.* Mathematical Association of America, Washington, DC, 2000.
13. Richeson, David S., *Euler's Gem: The Polyhedron Formula and the Birth of Topology.* Princeton University Press, Princeton, New Jersey, 2008.
14. Roberts, S., On the figures formed by the intercepts of a system of straight lines in a plane, and on analogous relations in space of three dimensions, *Proceedings of the London Mathematical Society* **19** (1899), 405–422.
15. Sloane, N. J. A., The On-Line Encyclopedia of Integer Sequences, *Notices of the American Mathematical Society* **50** (2003), 912–915.
16. Steiner, J., Eineige Gesetze über die Theilung der Ebene und des Raumes, *J. für die Reine und Angewandte Mathematik* **1** (1826), 349–364.
17. Wetzel, J. E., On the division of the plane by lines, *American Mathematical Monthly* **85** (1978), 647–656.
18. Zimmerman, S., Slicing space, *College Mathematics Journal* **32** (2001), 126–128.

1.11 Problems

1. Suppose that n straight cuts through a circular pizza form $P(n)$ pieces. Show that the number of pieces not touching the crust (boundary) of the pizza is

$$\binom{n}{0} - \binom{n}{1} + \binom{n}{2}.$$

2. Let $P(n)$ denote the maximum number of pizza pieces we can make with n cuts through a circular pizza and let $T(n) = P(0) + P(1) + P(2) + \cdots + P(n)$. Find a simple formula for $T(n)$.

3. The difference table for the sequence $a_0, a_1, a_2, \ldots$ is shown in Table 1.5. The row of third differences consists of 0's. Show that

$$a_n = a_0\binom{n}{0} + b_0\binom{n}{1} + c_0\binom{n}{2}.$$

Table 1.5: A difference table

sequence	a_0		a_1		a_3		a_4		a_5		a_6		·
first differences		b_0		b_1		b_2		b_3		b_4		·	
second differences			c_0		c_1		c_2		c_3		·		·
third differences				0		0		0		·		·	

4. Let $E(n)$ denote the maximum number of regions formed by n ellipses in the plane. Figure 1.13 shows that $E(5) \geq 42$. This problem discusses three proofs of the formula

$$E(n) = 2(n^2 - n + 1) \qquad \text{for } n = 1, 2, 3, \ldots.$$

Figure 1.13: Five ellipses and 42 regions

(a) Draw some ellipses to find $E(n)$ for $n = 1, 2, \ldots, 5$. Hint: Each pair of ellipses must intersect in four points.

(b) Justify the recurrence

$$E(n) = E(n-1) + 4(n-1) \qquad \text{for } n = 2, 3, 4, \ldots.$$

(c) Deduce the formula for $E(n)$ from the recurrence.

(d) Use a difference table to establish the formula for $E(n)$.

(e) Use Euler's formula to find $E(n)$.

5. Let $S(n) = 0^3 + 1^3 + 2^3 + \cdots + n^3$ for $n = 0, 1, \ldots$. Show that

$$S(n) = \frac{n^2(n+1)^2}{4},$$

say, by constructing a difference table.

6. Let $R(n)$ denote the maximum number of regions formed by n circles ($n \geq 1$) drawn on the surface of a sphere. This

problem outlines two proofs of the formula

$$R(n) = n^2 - n + 2.$$

(a) Verify that $R(1) = 2$, $R(2) = 4$, $R(3) = 8$, and $R(4) = 14$.

(b) Explain why $R(n) = 2P(n-1)$. Deduce the formula for $R(n)$.

(c) Suppose we have an optimal configuration of n circles. Show that each circle is partitioned into $2(n-1)$ arcs and that there are $2\binom{n}{2}$ intersection points.

(d) Deduce the formula for $R(n)$ from part (c) and Euler's formula for polyhedra.

7. Choose n points ($n \geq 1$) on a circle and form all chords joining them. Assume that no three chords pass through the same point inside the circle. Let $Q(n)$ denote the number of regions in the circle. Figure 1.14 shows that $Q(5) = 16$.

   (a) Verify that $Q(n) = 2^{n-1}$ for $n = 1, 2, \ldots, 5$.

   (b) Show that $Q(6) = 31$.

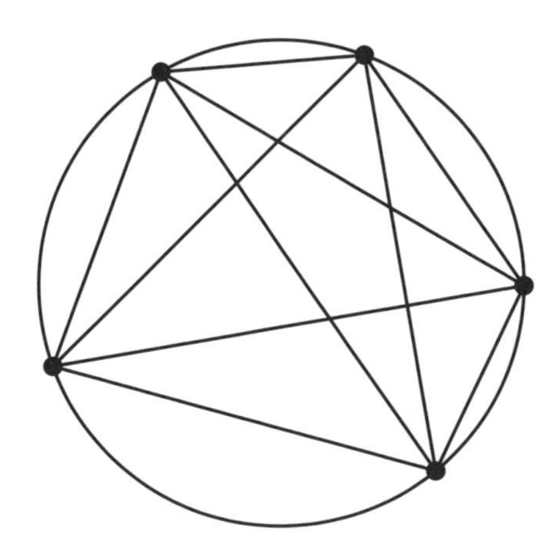

Figure 1.14: The chords determined by five points on a circle

(c) Explain why there are $\binom{n}{4}$ interior intersection points.

(d) Show that a suitable plane graph has $2\binom{n}{4} + n(n+1)/2$ edges.

(e) Use Euler's formula for plane graphs to show that

$$Q(n) = \binom{n}{0} + \binom{n}{2} + \binom{n}{4}.$$

8. Complete Figure 1.15 to produce a proof without words for the formula for $Q(n)$ in Problem 7.

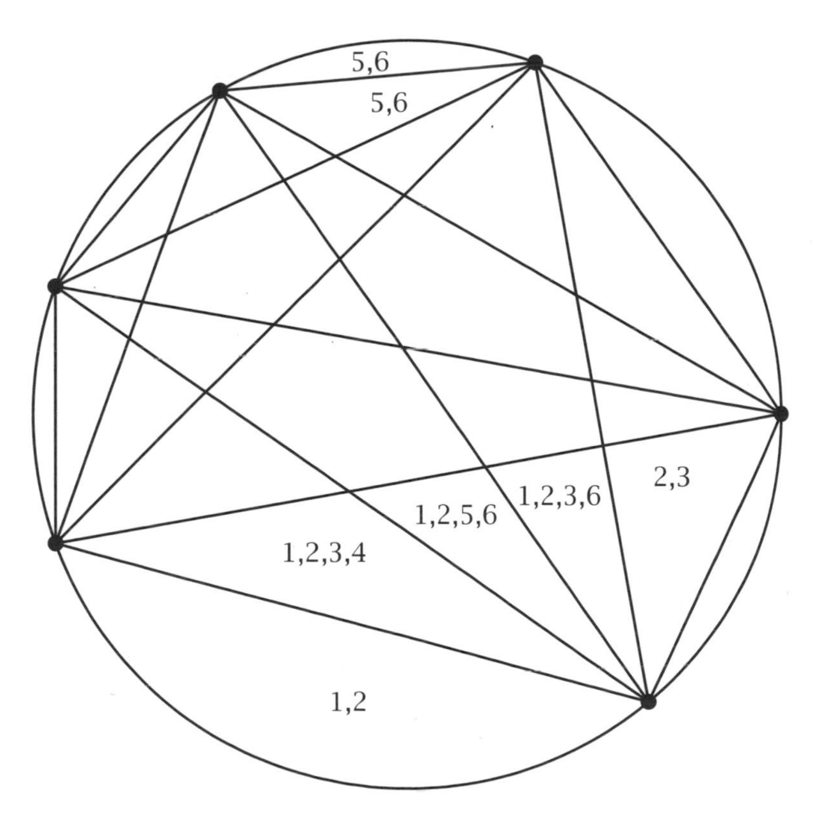

Figure 1.15: A proof without words: $Q(n) = \binom{n}{0} + \binom{n}{2} + \binom{n}{4}$

9. Explain why Figure 1.16 is a proof without words for the identity

$$1 + 2 + 3 + \cdots + n = \binom{n+1}{2}.$$

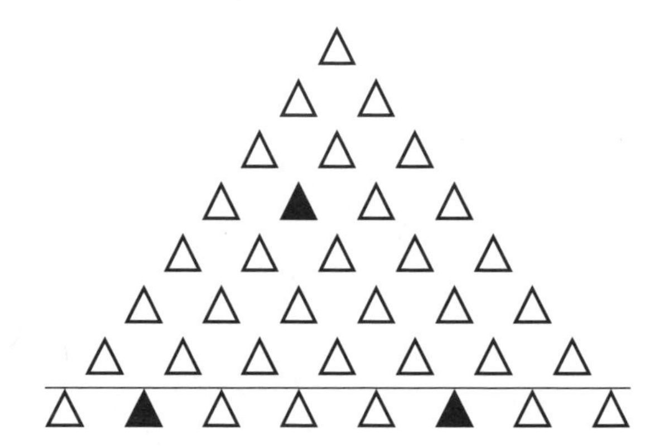

Figure 1.16: Proof without words: $1 + 2 + 3 + \cdots + n = \binom{n+1}{2}$

10. Produce a proof without words (similar to Figure 1.6) that the maximum number of pieces formed by n cuts through a circular pizza is

$$P(n) = 1 + \binom{n+1}{2}.$$

Hint: Use the labels $0, 1, \ldots, n$, and let the circle correspond to the label 0.

11. In the well-known *three utilities puzzle,* the object is to connect three houses to three utilities using noncrossing pipes in the plane. The attempt in Figure 1.17(a) has just one crossing. This problem guarantees that all such attempts are doomed to failure, even if we are allowed to move the houses and utilities to seemingly advantageous locations, as in Figure 1.17(b). Assume that the houses *can* be connected

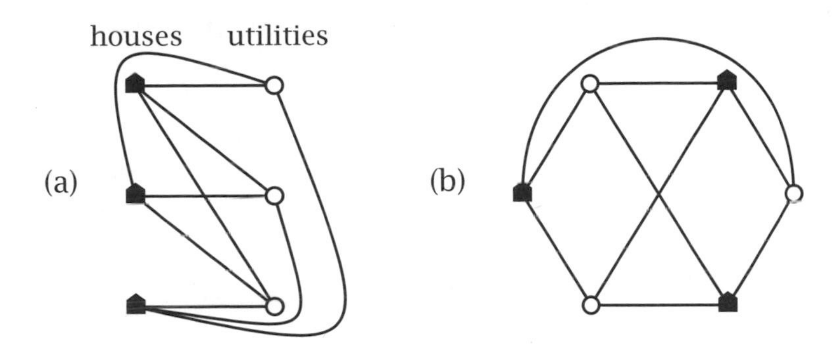

Figure 1.17: The three utilities puzzle

to the utilities, forming a connected plane graph with v vertices, e edges, and f faces.

(a) Explain why $4f \leq 2e$. Hint: Each face must have at least four edges.

(b) Deduce the value of f from the values $v = 6$ and $e = 9$.

(c) Use parts (a) and (b) to arrive at a contradiction.

12. Give an example of a polyhedron with the specified number of vertices and edges for each $n = 3, 4, 5, \ldots$.

(a) $2n$ vertices and $3n$ edges.

(b) $n + 1$ vertices and $2n$ edges.

(c) $n + 2$ vertices and $3n$ edges.

(d) $2n$ vertices and $4n$ edges.

(e) kn vertices and $(2k - 1)n$ edges for $k = 2, 3, \ldots$.

13. A polyhedron has v vertices and e edges. Every face is a triangle. Show that $e = 3v - 6$.

14. Show that three straight cuts through a circular pizza cannot produce seven pieces with equal areas. Hint: If the seven

pieces had equal areas, what fraction of the pizza would lie on one side of a cut?

15. Suppose that n cuts through a circular pizza form $P(n)$ pieces, where $n \geq 2$. Use Ismailescu's theorem to show that

$$\frac{\text{area of biggest piece}}{\text{area of smallest piece}} > \frac{n(n+1)}{8n-2}.$$

16. What does the On-Line Encyclopedia of Integer Sequences say about the sequence 2, 7, 1, 8, 2, 8, 1, 8? (These are the first eight digits of Euler's number e, the base for natural logarithms.)

17. Let $G(n)$ denote the maximum number of pieces of fruit formed by n plane slices through a spherical grapefruit. This problem outlines one derivation of the grapefruit-cutter's formula

$$G(n) = \binom{n}{0} + \binom{n}{1} + \binom{n}{2} + \binom{n}{3}.$$

(a) Verify that $G(0) = 1$, $G(1) = 2$, $G(2) = 4$, $G(3) = 8$, and $G(4) = 15$ directly.

(b) Justify the *grapefruit-cutter's recurrence,* valid for $n = 1, 2, \ldots$:

$$G(n) = G(n-1) + \binom{n-1}{0} + \binom{n-1}{1} + \binom{n-1}{2}.$$

Hint: Use Figure 1.18 as an inspiration.

(c) Use the recurrence in part (b) to show that the difference table for $G(n)$ contains only 0's in the row of fourth differences. See Table 1.6.

(d) Deduce the grapefruit-cutter's formula.

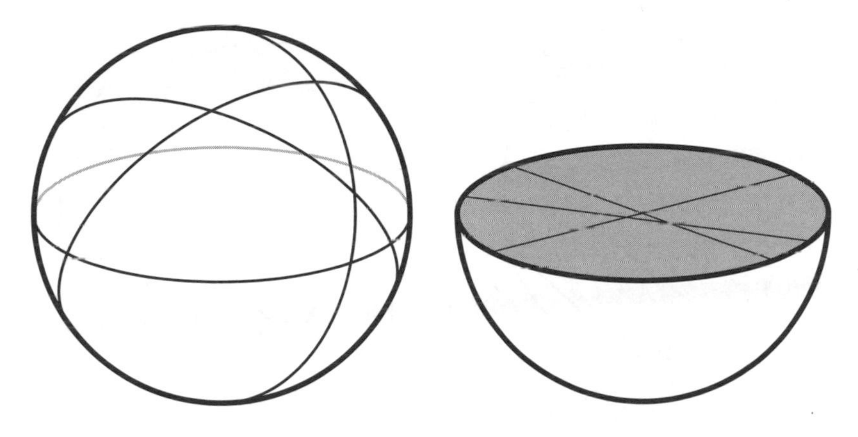

Figure 1.18: A recurrence for cutting a grapefruit

Table 1.6: A difference table for the grapefruit-cutter's formula

sequence	1		2		4		8		15		26		·
first differences		1		2		4		7		11		·	
second differences			1		2		3		4		·		·
third differences				1		1		1		·		·	
fourth differences					0		0		·		·		·

18. This problem outlines another proof of the Euler relation

$$p = e - v + 1$$

in the context of cutting pizzas. Start with an arrangement of n cuts through a circular pizza. The cuts need not intersect pairwise. Trim off the curved crust with a knife to form a polygonal pizza, as in Figure 1.19. Make diagonal cuts within each piece (the thin lines in the figure) so that all of the pizza pieces are triangles. The resulting configuration is a *triangulated polygon*. Let there be B boundary vertices, I interior vertices, E edges, and T triangles. The total number

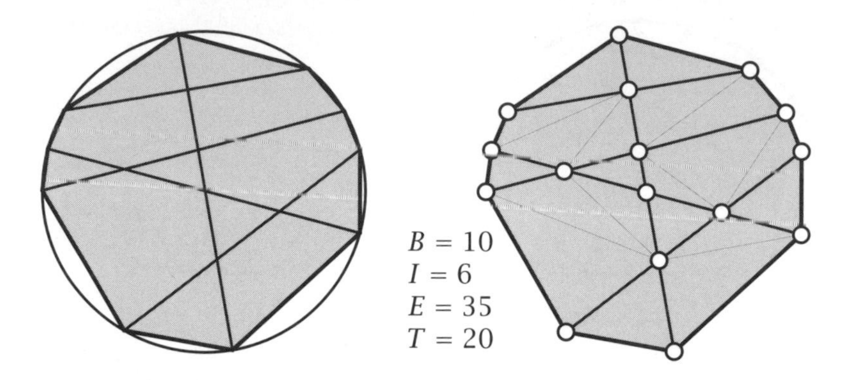

Figure 1.19: The pizza is trimmed and triangulated

of vertices is $V = I + B$.

(a) Explain why $2E = 3T + B$.

(b) In Chapter 2 (Section 2.7), we discuss the triangulated polygon theorem, which asserts that $T = 2I + B - 1$. Use this result to show that $E = 3I + 2B - 3$.

(c) Show that the triangulated polygon satisfies the Euler relation $T = E - V + 1$.

(d) Deduce that $p = e - v + 1$ holds for the original configuration. Hint: What happens if you remove one of the diagonals we inserted to triangulate the polygon?

There's a pizza place near where I live that sells only slices.
In the back you can see a guy tossing a triangle in the air.
STEVEN WRIGHT

Any good theorem should have several proofs,
the more the better.
SIR MICHAEL ATIYAH

2

Count on Pick's Formula

Questions are never indiscreet.
Answers sometimes are.
OSCAR WILDE

2.1 The Orchard and the Dollar

An area question. What is the area of the orchard in Figure 2.1 if the rows and columns of trees are 1 unit apart?

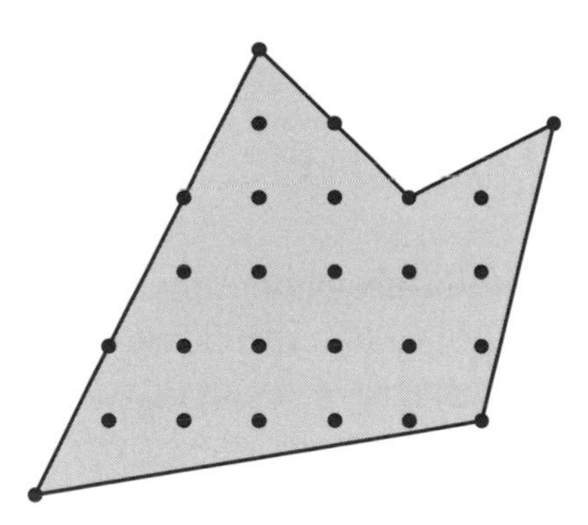

Figure 2.1: Find the area of the pentagonal orchard

Counting questions. How many ways are there to make change for a dollar from a supply of quarters, dimes, and nickels? What about change for D dollars?

Although the questions we have posed appear entirely unrelated, a remarkable formula discovered by the Austrian mathematician Georg Pick (1859–1942) can be used to answer both of them. Pick's formula relates problems from two entirely different worlds—counting problems from discrete mathematics and area problems from plane geometry, the quintessential branch of "indiscrete" (continuous) mathematics. Pick's formula lets us find the orchard's area by counting trees, and let us answer dollar-changing questions by computing areas of polygons.

Our goal is to understand Pick's formula and its applications. We first use several different methods to find the orchard's area and to count the ways to make change for a dollar. The solutions lead us naturally to Pick's formula, which we then state, verify in a few special cases, and apply to more general money-changing problems. We next give two proofs of Pick's formula, both relying on elementary geometric notions. Finally, we explore generalizations to higher dimensions.

2.2 The Area of the Orchard

While it is not immediately clear how to compute the exact area of the orchard, there is a quick way to find the *approximate* area. Merely count trees. Each of the orchard's 27 trees is the center of a 1-by-1 square, as shown in Figure 2.2. The total area of these unit squares is approximately the area of the orchard, and so

$$\text{area of orchard} \approx \text{number of trees} = 27.$$

(The notation $X \approx Y$ means X is approximately equal to Y.)

Closer scrutiny of Figure 2.2 yields a better approximation. If a tree is on the orchard's boundary, then about half of the surrounding unit square protrudes outside the or-

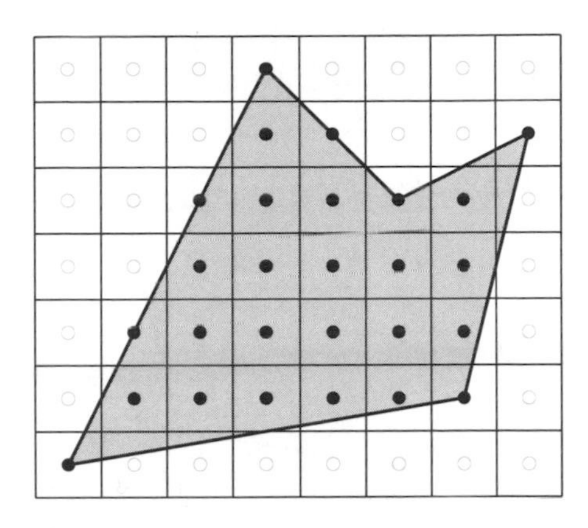

Figure 2.2: Each tree is the center of a unit square

chard. There are eight such boundary trees, and thus

$$\text{area of orchard} \approx \begin{array}{c}\text{number}\\ \text{of trees}\end{array} - \begin{array}{c}\text{half the number}\\ \text{of boundary trees}\end{array}$$

$$= 27 - \tfrac{1}{2}(8) = 23.$$

We will soon see that the exact area of the orchard is 22. So our tree-counting approximation is off by 1 unit of area.

Let us outline two standard geometric methods to compute the exact area of the orchard. It is convenient to introduce Cartesian coordinates and partition the pentagonal orchard into three triangles, as shown in Figure 2.3. We want to find the areas A_1, A_2, and A_3 of the triangles. Applying the familiar formula

$$\text{area of triangle} = \frac{\text{base} \times \text{height}}{2}$$

is awkward because none of the sides of the triangles is parallel to the x- or y-axes. Further geometric analysis is needed to find the bases and heights; we do not delve into the details. A second approach relies on Heron's formula, a more complicated and less well-known formula that gives

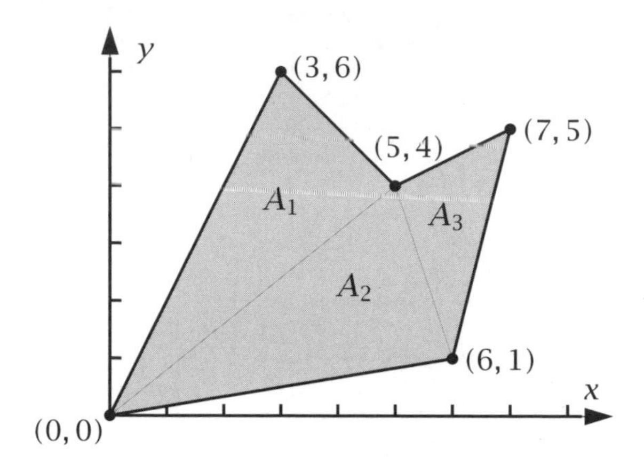

Figure 2.3: The pentagon is partitioned into three triangles

the area of a triangle in terms of its sides. The details are again somewhat messy. See Problem 3.

Getting Framed

Figure 2.4 illustrates a framing method that leads directly to the area of a triangle. The figure frames one of the orchard's triangles in a 6-by-4 rectangle. The areas of the three unshaded right triangles surrounding the shaded triangle are

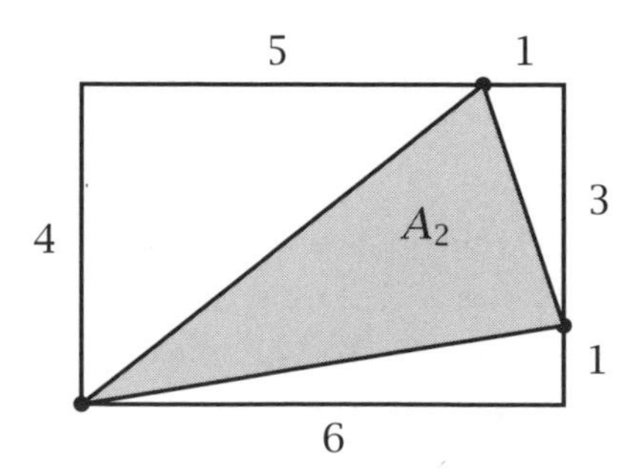

Figure 2.4: The area of the framed triangle is $A_2 = 19/2$

easy to find, and subtraction gives

$$A_2 = 6 \times 4 - \left(\frac{5 \times 4}{2} + \frac{1 \times 3}{2} + \frac{6 \times 1}{2} \right) = \frac{19}{2}.$$

We leave you to frame the other two triangles in the orchard and find $A_1 = 9$ and $A_3 = 7/2$. Thus the total area of the orchard is

$$A_1 + A_2 + A_3 = 9 + \frac{19}{2} + \frac{7}{2} = 22.$$

The framing idea works in general and shows that the triangle with vertices $(0,0)$, (x_1, y_1), and (x_2, y_2) satisfies

$$\text{area of triangle} = \frac{|x_2 y_1 - x_1 y_2|}{2}. \qquad (2.1)$$

This formula is best understood in the context of vectors and determinants, an interesting avenue we choose not to explore.

We now have a general strategy to find the area of any polygon in the xy-plane: Partition the polygon into triangles, apply (2.1) repeatedly, and sum the resulting areas.

2.3 Twenty-nine Ways to Change a Dollar

There are 29 ways to make change for a dollar from a supply of quarters, dimes, and nickels. We demonstrate this fact three different ways. Our solutions help us see that the dollar-changing problem is related to the area of a certain triangle.

Solution 1: List Them All

Our first solution is straightforward and uninspiring. We merely list all possibilities. We must select q quarters, d dimes, and n nickels so that

$$25q + 10d + 5n = 100.$$

Table 2.1: The 29 ways to make change for a dollar

dimes	$q = 0$	$q = 1$	$q = 2$	$q = 3$	$q = 4$
$d = 10$	$(0, 10, 0)$				
$d = 9$	$(0, 9, 2)$				
$d = 8$	$(0, 8, 4)$				
$d = 7$	$(0, 7, 6)$	$(1, 7, 1)$			
$d = 6$	$(0, 6, 8)$	$(1, 6, 3)$			
$d = 5$	$(0, 5, 10)$	$(1, 5, 5)$	$(2, 5, 0)$		
$d = 4$	$(0, 4, 12)$	$(1, 4, 7)$	$(2, 4, 2)$		
$d = 3$	$(0, 3, 14)$	$(1, 3, 9)$	$(2, 3, 4)$		
$d = 2$	$(0, 2, 16)$	$(1, 2, 11)$	$(2, 2, 6)$	$(3, 2, 1)$	
$d = 1$	$(0, 1, 18)$	$(1, 1, 13)$	$(2, 1, 8)$	$(3, 1, 3)$	
$d = 0$	$(0, 0, 20)$	$(1, 0, 15)$	$(2, 0, 10)$	$(3, 0, 5)$	$(4, 0, 0)$

quarters

A systematic approach accounts for each possibility once. Table 2.1 lists the 29 triples (q, d, n) of nonnegative integers that satisfy our equation. This method produces the correct answer with little effort but yields scant insight into the general problem of making change for D dollars.

Solution 2: Coins on the Table

Our second solution requires more imagination. We create a scenario that helps us keep track of the ways to make change. First, observe that once some quarters and dimes have been selected to make an amount not exceeding a dollar, there is no choice for the number of nickels to make up the balance. This observation allows us to ignore the nickels altogether. We focus on counting the ways to make *at most* one dollar from a supply of quarters and dimes.

Now place four quarters and ten dimes on a table. These

coins make two dollars and suffice to make any feasible amount not exceeding a dollar in all possible ways. Suppose we select q quarters and d dimes. There are five choices for q (namely, $q = 0, 1, \ldots, 4$) and eleven choices for d ($d = 0, 1, \ldots, 10$). Hence the total number of coin combinations available to us is $5 \times 11 = 55$. The amount of money we have selected is $25q + 10d$ cents, and the remaining $4 - q$ quarters and $10 - d$ dimes on the table account for the complementary $200 - (25q + 10d)$ cents. There are three ways for us to have exactly one dollar, namely, $(q, d) = (4, 0)$, $(2, 5)$, and $(0, 10)$. The 52 other coin combinations must occur as 26 complementary pairs, where one combination in each pair makes strictly less than a dollar, while the other makes strictly more than a dollar. It follows that there are $3 + 26 = 29$ ways to make an amount not exceeding a dollar with quarters and dimes, and therefore 29 ways to make change for a dollar using quarters, dimes, and nickels.

Solution 3: Integer Pairs

Our third solution involves counting special points in a triangle and will bring us to the brink of Pick's formula. We seek the number of triples of nonnegative integers (q, d, n) satisfying

$$25q + 10d + 5n = 100.$$

As before, we ignore the nickels and focus on the pairs (q, d) satisfying the inequalities

$$25q + 10d \leq 100, \qquad q \geq 0, \qquad d \geq 0.$$

In the Cartesian plane, these inequalities define the shaded right triangle in Figure 2.5.

Consider the *integer pairs* in[1] the triangle—points (q, d) whose coordinates q and d are both integers. Each integer

[1]When we refer to points "in" a polygon, we include the boundary as well as the interior.

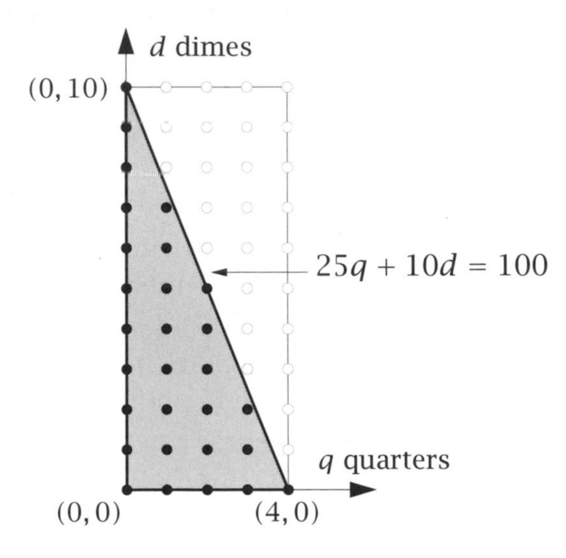

Figure 2.5: Integer pairs in a triangle and change for a dollar

pair in the shaded triangle corresponds to a way to make change for a dollar. For instance, $(q, d) = (2, 3)$ corresponds to two quarters, three dimes, and four nickels. We can count the integer pairs (q, d) in the shaded triangle directly to see that there are 29 ways to make change for a dollar.

The essential ideas of solutions 1 and 2 are contained within solution 3. Note that each of the 29 integer pairs in the triangle in Figure 2.5 corresponds to an entry in Table 2.1. Also, in the full rectangular array of $5 \times 11 = 55$ integer pairs in Figure 2.5, we can match the point (q, d) below the hypotenuse with the point $(4 - q, 10 - d)$ above the hypotenuse. The three integer pairs (q, d) on the hypotenuse correspond to the three ways to make a dollar using just quarters and dimes. So the coins-on-the-table approach is captured by the symmetry of two right triangles in Figure 2.5.

An Approximation

Figure 2.6 interprets our third solution to the dollar-changing question in a manner that resembles our approximation technique for the orchard's area. Each integer pair in the

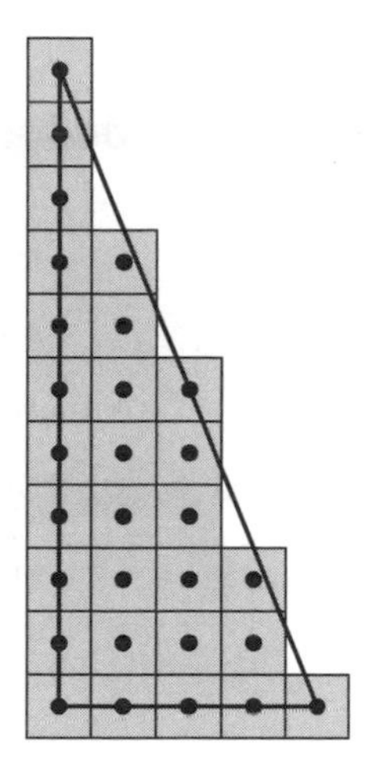

Figure 2.6: Integer pairs and unit squares

triangle is the center of a shaded unit square. The number of ways to make change for a dollar is thus the total area of the unit squares. But the triangle and shaded region have approximately the same area. The unit squares on the boundary protrude outside the triangle, and so

$$\begin{matrix}\text{area} \\ \text{of triangle}\end{matrix} \approx \begin{matrix}\text{number of} \\ \text{integer pairs}\end{matrix} - \begin{matrix}\text{half the number} \\ \text{of boundary pairs}\end{matrix}.$$

The area of the triangle is $(4 \times 10)/2 = 20$, and there are 16 integer pairs on the boundary. Therefore,

$$20 \approx \text{number of integer pairs} - \tfrac{1}{2}(16),$$

and we find that the number of ways to make change for a dollar is approximately 28. We know the exact answer is 29. Our approximation is again off by 1.

2.4 Lattice Polygons and Pick's Formula

We now state Pick's formula. The point (a, b) in the Cartesian plane is a *lattice point* provided the coordinates a and b are both integers. Each vertex in a *lattice polygon* is a lattice point. For a given lattice polygon, we let

$$\begin{aligned} L &= \text{the total number of lattice points} \\ &\quad\ \text{inside or on the boundary} \\ \text{and} \quad B &= \text{the number of boundary lattice points.} \end{aligned}$$

Our earlier work justifies the approximation

$$\text{area of polygon} \approx L - \tfrac{1}{2}B.$$

Pick's formula gives us the *exact* area of a lattice polygon in terms of L and B; our approximation is always 1 unit too big.

Pick's formula. If a simple lattice polygon has L lattice points and B boundary lattice points, then

$$\text{area of polygon} = L - \tfrac{1}{2}B - 1.$$

Example 1. The orchard in Figure 2.1 is a lattice polygon, and a direct count of the trees shows that $L = 27$ and $B = 8$. Pick's formula confirms our earlier result:

$$\text{area of orchard} = L - \tfrac{1}{2}B - 1 = 27 - \tfrac{1}{2}(8) - 1 = 22.$$

Pick's formula implies that the area of a lattice polygon is either an integer or half an odd integer, a somewhat surprising outcome given the seemingly arbitrary shapes for a lattice polygon.

Pick's formula is most frequently invoked to find the area of a lattice polygon by transforming the problem to one of counting lattice points. Our interest lies in applying the formula the other way around to answer counting questions. The idea is to recast a counting question as one involving lattice points in a polygon and then use Pick's formula to relate the number of lattice points to the area of the polygon. The polygon is often a triangle or a rectangle, whose area is easy to compute by other means.

Students and scholars accustomed to the standard methods of computing area using formulas from classical geometry and calculus are usually shocked by the presence of discrete parameters in Pick's area formula. The mixture of the discrete and the continuous accounts for the surprisingly recent vintage (1899) of a result rooted entirely in elementary mathematics.

Pick produced other elegant results across a spectrum of mathematical specialties. Early in his career, he was an assistant to the physicist Ernst Mach. Later he was instrumental in securing an academic position at the University of Prague for his good friend Albert Einstein. Pick met a tragic end, dying in 1942 at age 82 in Theresienstadt, a Nazi concentration camp.

Excluded Polygons

Pick's formula is valid for *simple* lattice polygons. The poly-

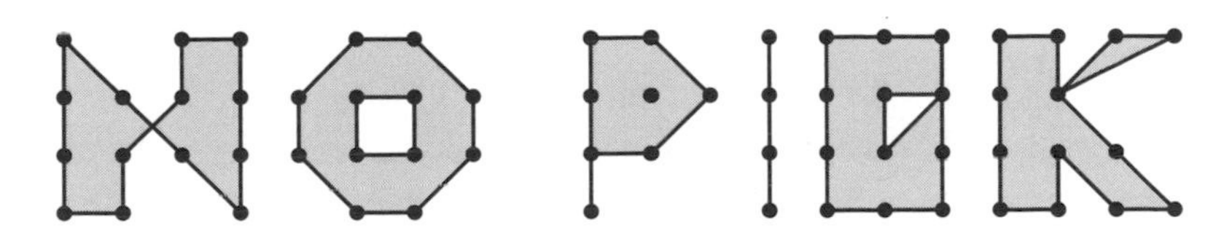

Figure 2.7: Six configurations not covered by Pick's formula

gon's boundary must be a sequence of segments forming a continuous loop with no crossings, no repeated vertices, and no holes. Figure 2.7 shows several configurations not covered by Pick's formula.

Rectangles and Right Triangles

We now explain why Pick's formula is true for *axis-parallel* rectangles (those with sides parallel to the x- and y-axes) and right triangles. Later we will see that the validity of the formula for all other lattice polygons hinges on these two special cases.

Example 2. An axis-parallel a-by-b lattice rectangle has $a + 1$ lattice points on each of the two horizontal boundary segments and $b + 1$ lattice points on each of the two vertical boundary segments. (Figure 2.8 shows the case $a = 9$ and $b = 6$.) Therefore, $L = (a + 1)(b + 1)$ and $B = 2a + 2b$, and Pick's expression is

$$L - \tfrac{1}{2}B - 1 = (a + 1)(b + 1) - \tfrac{1}{2}(2a + 2b) - 1 = ab,$$

which is indeed the area of the rectangle.

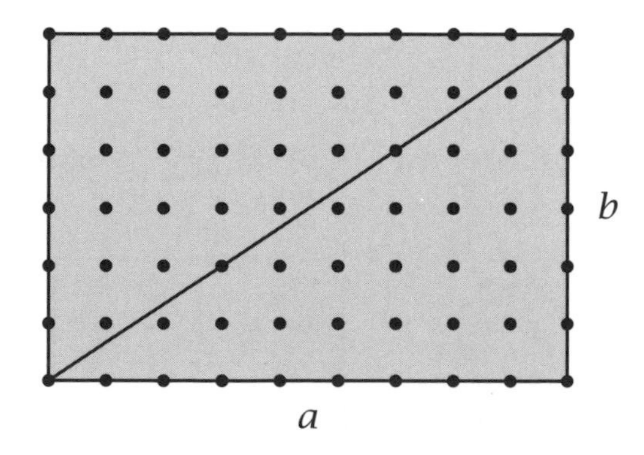

Figure 2.8: An a-by-b lattice rectangle bisected by a diagonal

Example 3. The diagonal in Figure 2.8 cuts the rectangle into two lattice right triangles with legs a and b. To confirm Pick's formula in this case, we must show that each right triangle satisfies

$$L - \tfrac{1}{2}B - 1 = \frac{ab}{2}. \tag{2.2}$$

Let B^* denote the number of lattice points on the diagonal, including the two endpoints. Then the number of boundary lattice points for each right triangle is

$$B = a + b + B^* - 1.$$

Because the B^* diagonal lattice points contribute to both right triangles, we have $2L = (a+1)(b+1) + B^*$. It follows that

$$L = \frac{(a+1)(b+1) + B^*}{2},$$

and it is now straightforward to verify that (2.2) is satisfied.

Example 3 would be essentially the same if we had selected the other diagonal of the rectangle, producing a pair of right triangles with different orientations. We will present our arguments for typical configurations with the understanding that the same reasoning applies to a whole class of essentially identical configurations.

We did not need to know the value of B^* in Example 3. But it is not difficult to show that

$$B^* = \gcd(a, b) + 1,$$

where $\gcd(a, b)$ is the *greatest common divisor* of a and b. For instance, when $a = 9$ and $b = 6$, we have

$$B^* = \gcd(9, 6) + 1 = 3 + 1 = 4.$$

The four lattice points on the segment joining $(0, 0)$ and $(9, 6)$ are $(0, 0)$, $(3, 2)$, $(6, 4)$, and $(9, 6)$, as depicted in Figure 2.8. More generally, the number of lattice points on the segment joining the lattice points (a_1, b_1) and (a_2, b_2) is

$$\gcd(a_2 - a_1, b_2 - b_1) + 1. \tag{2.3}$$

The reasoning in Example 3 gives a formula for the number of lattice points in an axis-parallel right triangle.

Proposition 1. The total number of lattice points in the lattice right triangle with vertices $(0,0)$, $(a,0)$, and $(0,b)$ is

$$L = \frac{(a+1)(b+1) + \gcd(a,b) + 1}{2}.$$

Challenge 1. Show that the number of lattice points in the rescaled lattice right triangle with vertices $(0,0)$, $(Na,0)$, and $(0,Nb)$ is given by

$$L = \left(\frac{ab}{2}\right)N^2 + \left(\frac{a+b+\gcd(a,b)}{2}\right)N + 1.$$

We mention this formula again in Section 2.9.

2.5 Making Change

We are ready to count the ways to make change for D dollars. We could modify our earlier coins-on-the-table method, but it is easier at this point to let Proposition 1 do the heavy lifting for us.

We seek the number of solutions to the equation

$$25q + 10d + 5n = 100D,$$

where q, d, and n denote the number of quarters, dimes, and nickels. We divide by 5 and obtain

$$5q + 2d + n = 20D.$$

Once q and d are specified, n is determined, and so we examine solutions to the inequalities

$$5q + 2d \leq 20D, \qquad q \geq 0, \qquad d \geq 0. \tag{2.4}$$

The three inequalities determine a lattice right triangle with vertices $(0,0)$, $(4D,0)$, and $(0,10D)$. Each lattice point in the triangle corresponds to a way to make change for D dollars. Apply Proposition 1 with $a = 4D$ and $b = 10D$ to see that the number of lattice points is

$$L = \frac{(4D+1)(10D+1) + \gcd(4D,10D) + 1}{2}.$$

Because $\gcd(4D,10D) = 2D$, we have $L = 20D^2 + 8D + 1$.

The lattice points in the interior of the triangle correspond to ways to make change for D dollars using at least one coin of each of the three denominations. It is not difficult to show that there are $16D$ lattice points on the boundary of the triangle, and thus there are $20D^2 - 8D + 1$ ways to make change in this case. The following theorem summarizes our work and answers the dollar-changing question at the start of the chapter.

Dollar-changing theorem. The number of ways to make change for D dollars from a supply of quarters, dimes, and nickels is

$$20D^2 + 8D + 1.$$

Also, the number of ways to make change using at least one quarter, one dime, and one nickel is

$$20D^2 - 8D + 1.$$

The two formulas in the dollar-changing theorem are identical except for the sign of the middle term. This is not a coincidence and will be explained in Section 2.9.

The same reasoning used for the dollar-changing theorem leads to the following general result.

Three denominations theorem. With coins of three denominations d_1, d_2, and 1, the number of ways to make d_1d_2N units of money is

$$\left(\frac{d_1d_2}{2}\right)N^2 + \left(\frac{d_1 + d_2 + \gcd(d_1, d_2)}{2}\right)N + 1$$

for $N = 1, 2, \ldots$.

We will see ways to answer questions involving four denominations later. As a preview, try your hand at the following problem.

Challenge 2. How many ways are there to make change for a dollar if quarters, dimes, nickels, and pennies are available?

2.6 Pick's Formula: First Proof

It is time to show that Pick's formula is valid for all lattice polygons. We will give two proofs, each with its own virtues. Our first proof argues that Pick's formula is true for the smallest lattice triangles (those with exactly three lattice points) and that it remains true when such triangles are assembled to form more complicated polygons. Much of the work was already done in our discussion of axis-parallel rectangles and right triangles.

The Whole Is the Sum of Its Picks

Suppose that the simple lattice polygon P is partitioned into two simple lattice polygons P' and P'', as in Figure 2.9. Of course, the areas satisfy

$$\text{area of } P = \text{area of } P' + \text{area of } P''.$$

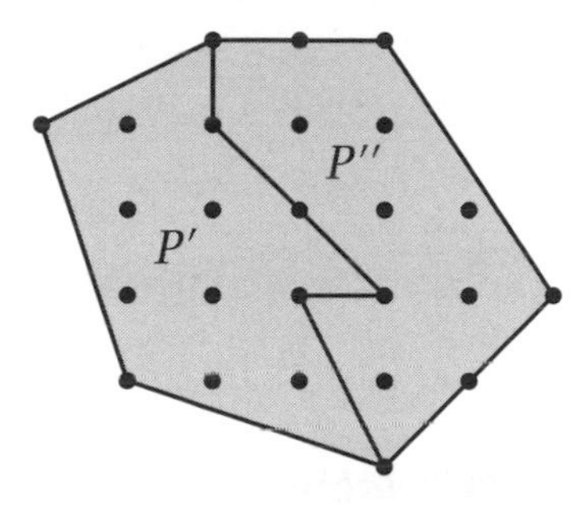

Figure 2.9: Area is additive

Now assume that

$$\begin{aligned} \text{area of } P' &= L' - \tfrac{1}{2} B' - 1 \\ \text{and} \qquad \text{area of } P'' &= L'' - \tfrac{1}{2} B'' - 1 \end{aligned}$$

in accordance with Pick's formula. We will show that the big polygon P also satisfies Pick's formula. Let the common boundary of the two smaller polygons P' and P'' contain B^* lattice points, including the two endpoints. The polygon P satisfies

$$\begin{aligned} L &= L' + L'' - B^*, \\ \text{and} \qquad B &= B' + B'' - 2B^* + 2. \end{aligned}$$

Therefore,

$$\begin{aligned} \text{area of } P &= \text{area of } P' + \text{area of } P'' \\ &= \left(L' - \tfrac{1}{2} B' - 1\right) + \left(L'' - \tfrac{1}{2} B'' - 1\right) \\ &= (L' + L'' - B^*) - \tfrac{1}{2}(B' + B'' - 2B^* + 2) - 1 \\ &= L - \tfrac{1}{2} B - 1. \end{aligned}$$

The preceding computation shows that if P' and P'' satisfy the Pick relation, then so does P. This means that the Pick expression $L - \frac{1}{2}B - 1$ is *additive*, just like area. There is a subtractive version, too: If Pick's formula holds for P and P', then it must also hold for P''.

Repeated use of additivity implies that Pick's formula holds for any simple polygon that can be partitioned into axis-parallel rectangles and right triangles.

Primitive Lattice Triangles

We now analyze the "atomic" lattice polygons—the simplest ones from which all other lattice polygons are built. A lattice triangle is *primitive* provided its only lattice points are its three vertices. A primitive lattice triangle satisfies $L = B = 3$. By Pick's formula

$$\text{area of a primitive lattice triangle} = 3 - \frac{3}{2} - 1 = \frac{1}{2}.$$

We need to establish this fact independently. Indeed, the validity of Pick's formula in general reduces to this special case.

A *primitive lattice triangulation* is a partition of a lattice polygon into primitive lattice triangles. Figure 2.10 shows a primitive lattice triangulation of our orchard into 44 primitive lattice triangles.

The following theorem contains the two most important facts about primitive lattice triangles.

Primitive lattice triangles theorem.

(a) Evert lattice polygon has a primitive lattice triangulation.
(b) The area of every primitive lattice triangle is 1/2.

Part (a) implies that every lattice polygon can be constructed by adjoining primitive lattice triangles side-to-side

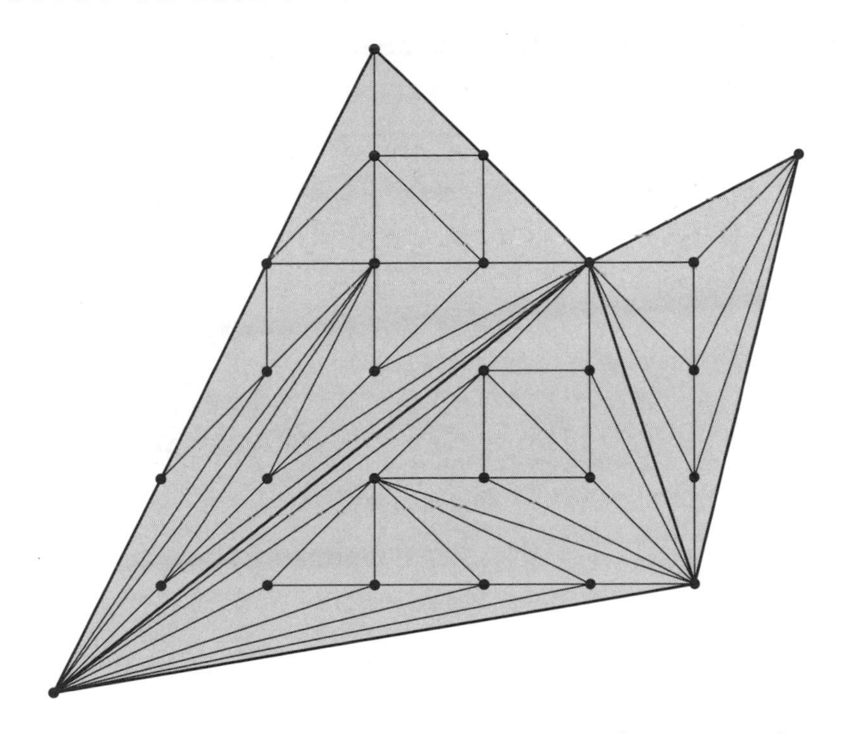

Figure 2.10: A triangulation into 44 primitive lattice triangles

one at a time. Part (b) and additivity assure us that Pick's formula holds as each triangle is added. Therefore, the theorem implies Pick's formula.[2]

The rest of this section explains why (a) and (b) are true.

Triangulations

Here is why (a) is true. First, any lattice polygon can be partitioned into (lattice) triangles by inserting diagonals. (This fact also arises when we triangulate art galleries in Chapter 3.) Second, any nonprimitive lattice triangle can be par-

[2]We are glossing over a technical issue about the order in which the primitive lattice triangles are adjoined. Some care is required to guarantee that a simple polygon occurs at each step.

titioned into two or three smaller lattice triangles by the following two operations:

- Insert edges joining an interior lattice point of a triangle to the three vertices.
- Insert an edge joining a boundary lattice point of a triangle to the opposite vertex.

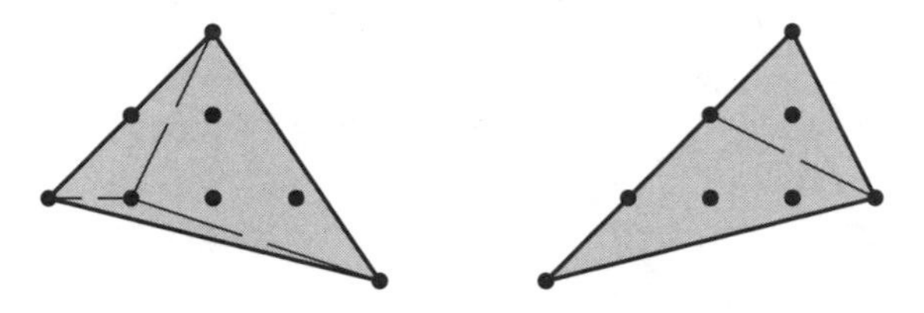

Figure 2.11: Insert edges in a nonprimitive lattice triangle

See Figure 2.11. Repeating these operations eventually leads to a primitive lattice triangulation. A lattice polygon usually has many primitive lattice triangulations; any one of these will do for our purposes.

Primitive Lattice Triangles: Framed Again

Here is why part (b) of the primitive lattice triangles theorem is true. Observe that a lattice triangle can be framed by an axis-parallel lattice rectangle in one of the two ways shown

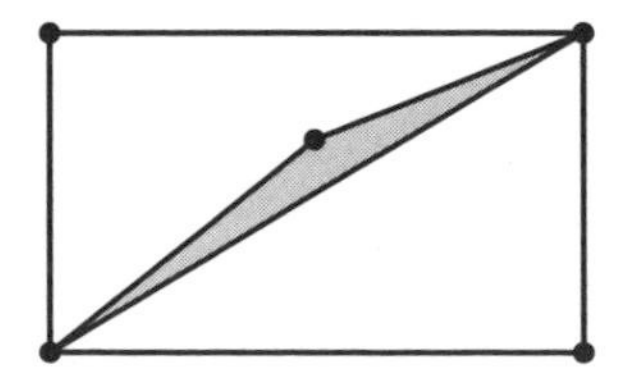
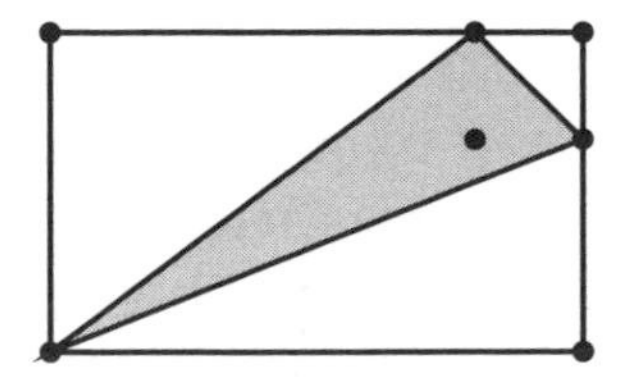

Figure 2.12: Two ways to frame a lattice triangle

in Figure 2.12. The configuration on the left applies to any primitive lattice triangle since the triangle on the right must contain an interior lattice point, as shown. The longest side of the primitive lattice triangle is a diagonal of the framing rectangle, and the only lattice points on this diagonal are the two endpoints.

When a primitive lattice triangle OUZ is framed by a rectangle, a lattice quadrilateral $OWUZ$ is formed, as in Figure 2.13. The quadrilateral satisfies Pick's formula because

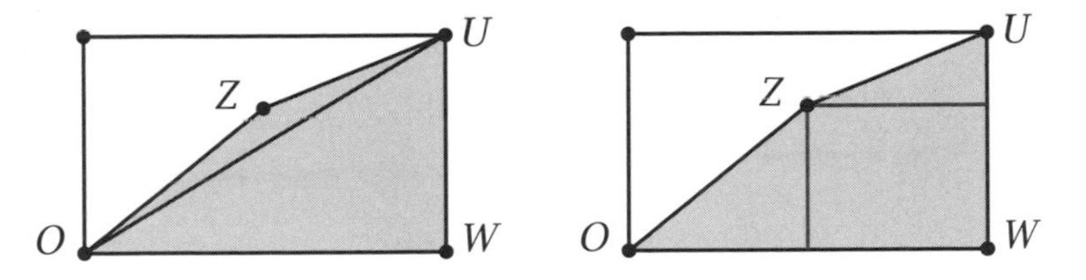

Figure 2.13: A primitive lattice triangle defines a quadrilateral

it can be partitioned into a lattice rectangle and two lattice right triangles. We also know that Pick's formula holds for the right triangle OWU. By subtraction, Pick's formula must hold for the primitive lattice triangle OUZ. Triangle OUZ satisfies $L = B = 3$, and Pick's formula tells us the area is $1/2$.

This completes our proof of Pick's formula. The key elements of the argument are additivity and the fact that the area of every primitive lattice triangle is $1/2$.

2.7 Pick's Formula: Second Proof

Another way to establish Pick's formula relies on counting. The tale is told of a shepherd who could glance across any large herd of sheep and immediately announce the number of sheep to astonished friends. When asked for his secret, he confessed that he merely counted the legs and divided by 4. The shepherd declined to reveal his leg-counting method.

Our second proof of Pick's formula uses the same strategy as the shepherd. To determine the area of a lattice polygon, we merely count the primitive lattice triangles in a triangulation and divide by 2. For instance, the triangulation of the orchard in Figure 2.10 contains 44 primitive lattice triangles, each with area 1/2. Therefore, the area of the orchard is $44/2 = 22$. To prove that this method always gives the right answer, we must show that any primitive lattice triangulation satisfies

$$\text{number of primitive triangles} \;=\; 2L - B - 2.$$

Division by 2 then shows that the area of the lattice polygon is $L - \frac{1}{2}B - 1$, which is Pick's formula.

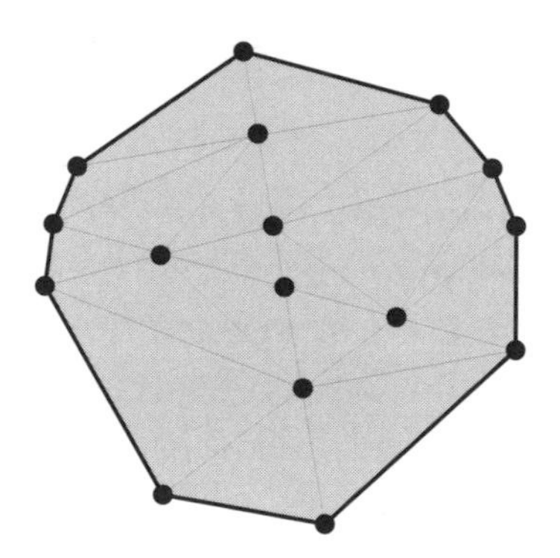

Figure 2.14: A triangulation of a polygon

Surprisingly, the above formula for the number of triangles does not require the vertices to be lattice points. Figure 2.14 shows a *triangulation* of a (not necessarily lattice) polygon. Two triangles are either disjoint, share a single vertex, or an entire side. Let L denote the total number of vertices in a triangulation, B the number of boundary vertices, and $I = L - B$ the number of interior vertices. Also, let T denote the number of triangles. For example, in Figure 2.14, we have $L = 16$, $B = 10$, $I = 6$, and $T = 20$. There is a basic relationship among B, I, and T that holds in any triangulation of a polygon.

Triangulated polygon theorem. If a triangulated polygon has B boundary vertices, I interior vertices, and T triangles, then $T = 2I + B - 2$.

The theorem implies that the number of primitive lattice triangles is

$$T = 2I + B - 2 = 2(I + B) - B - 2 = 2L - B - 2,$$

which completes our second proof of Pick's formula.

To see why the triangulated polygon theorem is true, consider the sum S of the angles of all T triangles in the triangulated polygon (Figure 2.15). We will compute S two different ways and equate the answers. First, the angles in each of the T triangles sum to $180°$. Thus

$$S = 180° T.$$

Second, each of the I interior vertices contributes $360°$, and

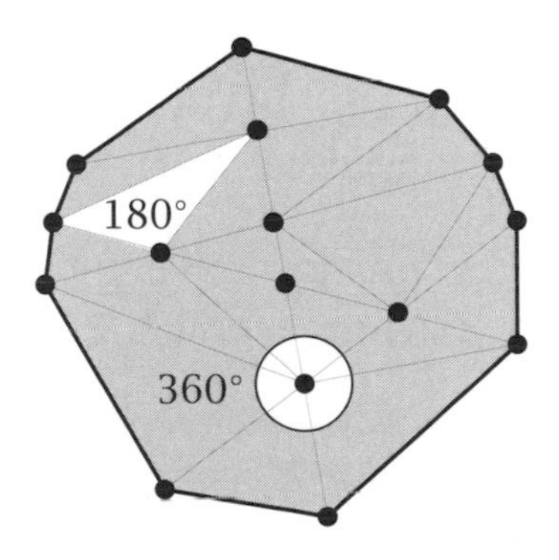

Figure 2.15: A proof of the triangulated polygon theorem

the sum of the angles at the B boundary vertices is $180°(B - 2)$ by a basic result in geometry. Thus

$$S = 360°I + 180°(B - 2).$$

We equate our two expressions for S to obtain

$$180°T = 360°I + 180°(B - 2)$$

and divide by 180° to conclude that $T = 2I + B - 2$.

By the way, the triangulated polygon theorem is also closely related to Euler's formula for plane graphs. See Problem 18 of Chapter 1 for details.

2.8 Batting Averages and Lattice Points

A baseball player's batting average is defined as

$$\text{batting average} = \frac{\text{number of hits}}{\text{number of at-bats}},$$

rounded to the nearest .001. For instance, a player with 3 hits in 11 at-bats has a batting average of .273 because

$$\frac{3}{11} = .272727\ldots \approx .273.$$

The fractions $41/150$ and $131/480$ also round to .273, as does $273/1000$, of course.

A baseball question. In how many ways can a baseball player achieve the batting average .273 with at most 2000 at-bats?

In more prosaic terms, we seek the number of fractions y/x satisfying the inequalities

$$0.2725 \le \frac{y}{x} < 0.2735 \qquad \text{and} \qquad 0 < x \le 2000.$$

With patience (or a computer), we could make a list of all such fractions to answer the question. Pick's formula gives a quick solution with less work.

A player with y hits in x at-bats has unrounded batting average y/x, which is the slope of the segment joining the origin $(0,0)$ and the lattice point (x,y). The batting average rounds to .273 when the slope satisfies

$$0.2725 = \frac{545}{2000} \le \frac{y}{x} < \frac{547}{2000} = 0.2735.$$

It follows that the batting averages we want to count are related to the lattice points in the triangle with vertices

$$(0,0),\ (2000,545),\quad \text{and}\quad (2000,547).$$

Figure 2.16 depicts this highly acute triangle in a somewhat distorted manner to emphasize the relevant features. The triangle has area 2000 since its vertical base is 2, and the height to that base is 2000. Three applications of formula (2.3) show that there are $B = 8$ boundary lattice points. According to Pick's formula,

$$2000 = L - \frac{8}{2} - 1 = L - 5,$$

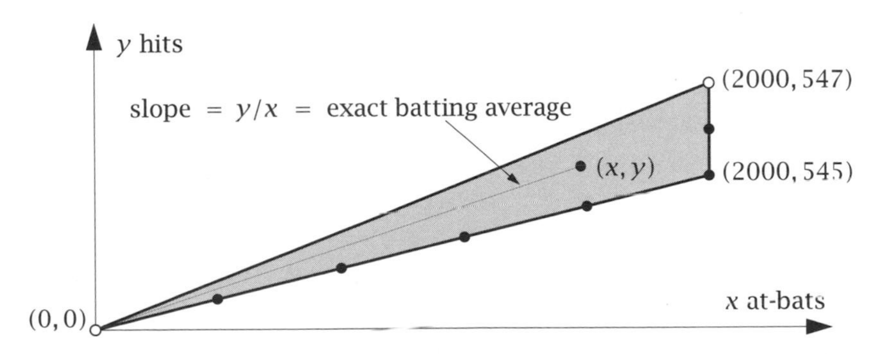

Figure 2.16: Lattice points and batting averages rounding to .273

and so there are $L = 2005$ lattice points in the triangle. Except for the two vertices $(0,0)$ and $(2000,547)$ shown as open dots, each of the lattice points corresponds to an exact batting average y/x that rounds to .273. We have shown that there are 2003 ways to achieve the batting average .273 with at most 2000 at-bats.

2.9 Three Dimensions and N-largements

There are two likely reasons Pick's formula attracted little attention for over half a century. First, interesting applications of the formula are not easy to find. The connections we have made between lattice points and counting questions were not in fashion in the early twentieth century. Second, Pick's formula does not have a natural extension to three dimensions.

Each vertex in a *lattice polyhedron* is a lattice point in three dimensions—a point (a,b,c), whose coordinates a, b, and c are integers. A three-dimensional version of Pick's formula would presumably express the volume of a lattice polyhedron in terms of the number of lattice points and boundary lattice points of the polyhedron. All hopes for such a formula are dashed by the following example, which describes two polyhedra with unequal volumes, but identical interior and boundary lattice point counts.

Example 4. Figure 2.17 shows a unit cube partitioned into five lattice polyhedra. The central tetrahedron has vertices

$$(0,0,1),\ (1,0,0),\ (0,1,0),\quad \text{and} \quad (1,1,1).$$

The four surrounding polyhedra are congruent right pyramids. Each of our five polyhedra has four lattice points, all of which fall on the boundary. Therefore, a naive Pick-like

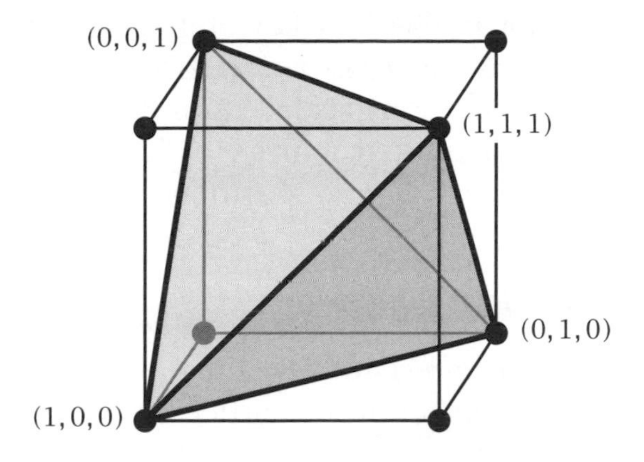

Figure 2.17: A unit cube partitioned into five tetrahedra

formula would imply that all five polyhedra have the same volume. But the formula

$$\text{volume of pyramid} = \frac{\text{area of base} \times \text{height}}{3}$$

shows that the volume of each of the four right pyramids is $1/6$, and since the volume of the whole cube is 1, the volume of the central tetrahedron is $1 - 4\,(1/6) = 1/3$.

There is also a theoretical reason Pick's formula does not extend neatly to three dimensions. Our proofs of Pick's formula rely on triangulations—partitions of a polygon into disjoint triangles by means of segments joining pairs of vertices. Similar partitions need not exist in three dimensions. Some polyhedra cannot be "tetrahedralized," that is, partitioned into tetrahedra by means of triangles through three vertices. A complicated example of such a polyhedron occurs in our discussion of art galleries in Chapter 3. See Figure 3.19.

Around 1960, several researchers looked at lattice point enumeration problems from a fresh viewpoint—a viewpoint that lets us generalize a modified version of Pick's formula to three dimensions. The first key step is to ask the right question: How does the number of lattice points change as a polygon or polyhedron is scaled up in size? The second step is to restrict attention to convex[3] polyhedra—which are guaranteed to have tetrahedral partitions.

N-largements: Two Dimensions

We first explore the effects of rescaling in two dimensions. Suppose a lattice polygon has L lattice points, B boundary lattice points, and area A. Rearranging Pick's formula gives

$$L = A + \tfrac{1}{2}B + 1.$$

If we rescale the polygon by changing its size while preserving its shape, then the lattice point counts and area change in a predictable manner, and the above relation remains valid.

Let P be a lattice polygon and let N be any nonnegative integer. Let P_N denote the rescaled lattice polygon obtained from P by replacing each point (x, y) with (Nx, Ny). See Figure 2.18. We think of enlarging P by a factor of N to form P_N, and refer to P_N as the *N-largement* of P. If $N = 1$, then P_N is P itself. If $N = 0$, then P_N collapses to the single lattice point $(0, 0)$.

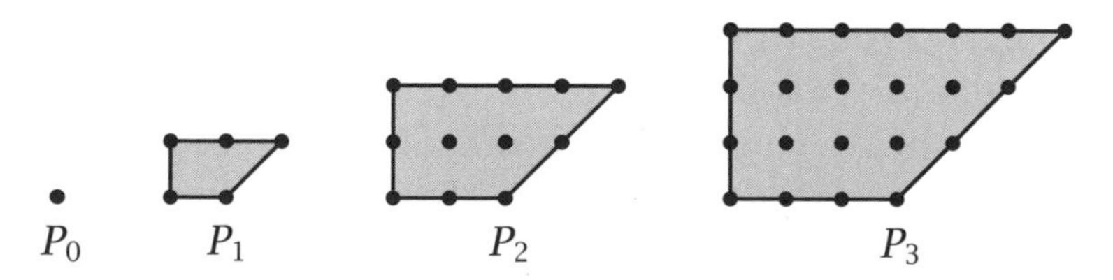

Figure 2.18: N-largements of a lattice polgyon

[3]A polyhedron is convex provided the line segment joining any two points of the polyhedron does not pass outside the polyhedron.

We introduce the parameters

$$\begin{aligned} L_N &= \text{the total number of lattice points inside or} \\ &\quad\ \text{on the boundary of the } N\text{-largement } P_N, \\ B_N &= \text{the number of boundary lattice points of } P_N, \\ A_N &= \text{the area of } P_N. \end{aligned}$$

The function L_N is called the *lattice point enumerator* for the polygon P. Our goal is to find a formula for L_N. We know that

$$L_N = A_N + \tfrac{1}{2} B_N + 1.$$

It is not difficult to show that the rescaling process gives

$$A_N = AN^2 \qquad \text{and} \qquad B_N = BN,$$

and it follows that the lattice point enumerator for P equals

$$L_N = AN^2 + \left(\tfrac{B}{2}\right) N + 1.$$

Pick's formula is the case $N = 1$. The formula for rescaled right triangles in Challenge 1 in Section 2.4 is another special case.

Although we have not defined the N-largement for negative values of N, nothing prevents us from setting $N = -1$ in the previous formula. We find that

$$L_{-1} = A - \tfrac{B}{2} + 1 = \left(A + \tfrac{B}{2} + 1\right) - B = L - B,$$

which is the number of interior lattice points of P. A similar computation shows that L_{-N} equals the number of interior lattice points of the N-largement P_N (except in the degenerate case $N = 0$).

The preceding discussion has brought us to the modified version of Pick's formula that can be generalized to three dimensions.

Lattice point enumerator theorem (two dimensions). The lattice point enumerator for the lattice polygon P is a quadratic polynomial

$$L_N = c_2 N^2 + c_1 N + c_0, \tag{2.5}$$

where c_2 is the area of P, c_1 is half the number of boundary lattice points, and $c_0 = 1$. Also, L_{-N} counts the number of lattice points in the interior of P_N for $N = 1, 2, \ldots$.

N-largements: Three Dimensions

Now let us look at rescaling in three dimensions. The lattice point enumerator of a polyhedron P counts lattice points in the N-largements of P.

Example 5. Figure 2.19 shows the unit cube P and some of its N-largements. Because the N-largement P_N has $N + 1$ lattice points along each edge, the lattice point enumerator of P is

$$L_N = (N + 1)^3 = N^3 + 3N^2 + 3N + 1.$$

Note that the coefficient of N^3 is 1, which is the volume of P, and the constant term is 1. We claim that there are

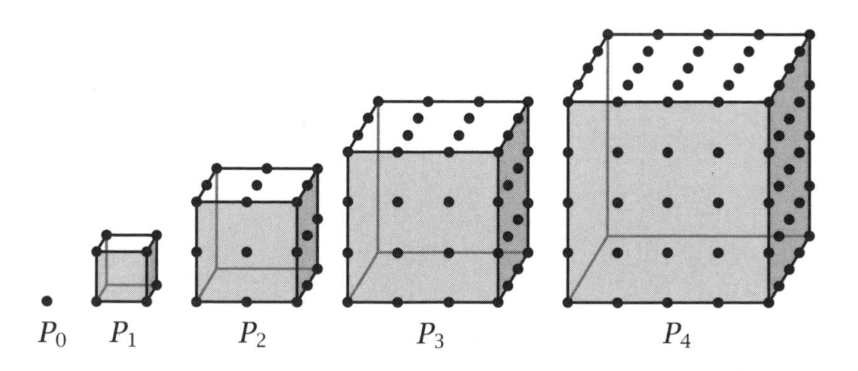

Figure 2.19: N-largements of the unit cube

$(N-1)^3$ lattice points in the interior of P_N. To see why, peel off the boundary lattice points from P_N to expose an inner cube with $N-1$ lattice points on each edge. Note that

$$L_{-1} = -N^3 + 3N^2 - 3N + 1 = -(N-1)^3.$$

The preceding example illustrates the three-dimensional analogue of formula (2.5).

Lattice point enumerator theorem (three dimensions). The lattice point enumerator for the convex lattice polyhedron P is a cubic polynomial

$$L_N = c_3 N^3 + c_2 N^2 + c_1 N + c_0, \tag{2.6}$$

where c_3 is the volume of P, and $c_0 = 1$. Also, $-L_{-N}$ counts the number of lattice points in the interior of the N-largement P_N for $N = 1, 2, \ldots$.

The coefficients of the lattice point enumerator can often be found by analyzing the polyhedron P. To illustrate the method, we find a formula for the number of ways to make change for $50N$ cents with a supply of quarters, dimes, nickels, and pennies. In other words, we find the number of solutions to the equation

$$25q + 10d + 5n + p = 50N \tag{2.7}$$

in nonnegative integers q, d, n, and p. The pennies can be ignored if we focus on making at most $50N$ cents with the other coins. Each solution of (2.7) corresponds to a lattice point (q, d, n) in the polyhedron defined by the inequalities

$$5q + 2d + n \le 10N, \qquad q \ge 0, \quad d \ge 0, \quad n \ge 0.$$

The polyhedron is an N-largement of the lattice polyhedron P with vertices

$$(0,0,0), \ (2,0,0), \ (0,5,0), \quad \text{and} \quad (0,0,10)$$

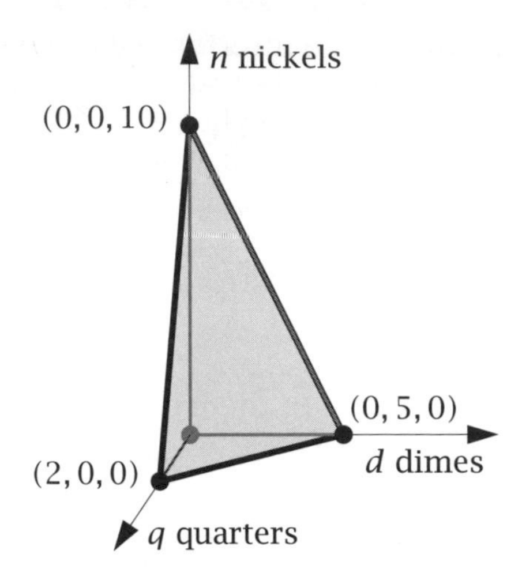

Figure 2.20: A polyhedron for making change

shown in Figure 2.20. The volume of P is $50/3$, and it is not difficult to verify that P has $L = 49$ lattice points and $L_{-1} = 2$ interior lattice points. So the lattice point enumerator

$$L_N = c_3 N^3 + c_2 N^2 + c_1 N + c_0$$

satisfies $c_3 = 50/3$ and $c_0 = 1$. Also,

$$\begin{aligned} 49 &= L_1 = c_3 + c_2 + c_1 + c_0 \\ \text{and} \quad 2 &= -L_{-1} = c_3 - c_2 + c_1 - c_0. \end{aligned}$$

A little algebra shows that $c_2 = 45/2$ and $c_1 = 53/6$. We have established the following result.

Theorem. The number of ways to make $50N$ cents from a supply of quarters, dimes, nickels, and pennies is

$$L_N = \left(\frac{50}{3}\right) N^3 + \left(\frac{45}{2}\right) N^2 + \left(\frac{53}{6}\right) N + 1 \qquad \text{for } N = 1, 2, \ldots.$$

In particular, the number of ways to make change for a dollar is $L_2 = 242$. This answers Challenge 2.

Higher Dimensions

For a d-dimensional lattice polyhedron P the lattice point enumerator L_N is a polynomial of degree d. The leading coefficient is the d-dimensional volume of P, the constant term is 1, and $(-1)^d L_{-N}$ is the number of interior lattice points of the N-largement P_N. Although it is difficult to visualize a d-dimensional polyhedron, the polynomial L_N can nonetheless be used to answer money-changing problems involving more than four denominations, as well as other enumerative questions.

2.10 Notes and References

The popular books by Ball [1] and Stewart [19] both contain chapters on Pick's formula, while the recent text by Beck and Robins [2] discusses Pick's formula and its generalizations in the broader context of modern discrete mathematics. The connection between Pick's formula and Euler's formula for plane graphs is discussed in [5, 9, 13]. Problem 10 is adapted from [8].

Pick's original paper is [16]. The original sources for the lattice point enumerator theorem are [6, 7, 14, 17]. The lattice point enumerator is more frequently called the Ehrhart polynomial.

1. Ball, Keith, *Strange Curves, Counting Rabbits, and Other Mathematical Explorations.* Princeton University Press, Princeton, New Jersey, 2003.
2. Beck, Matthias, and Robins, Sinai, *Computing the Continuous Discretely: Integer-Point Enumeration in Polyhedra.* Springer, New York, 2006.
3. Blatter, C., Another proof of Pick's area theorem, *Mathematics Magazine* **70** (1974), 200.
4. Bruckenheimer, M., and Arcavi, A., Farey series and Pick's area theorem, *Mathematical Intelligencer* **17** (1995), 200.
5. DeTemple, D., and Robertson, J. M., The equivalence of Euler's and Pick's theorems, *Mathematics Teacher* **67** (1974), 222–226.
6. Ehrhart, E., Sur les polyèdres rationnels homothétiques à *n* dimensions, *Comptes Rendus de l'Academie des Science* **254** (1962), 616–618.
7. Ehrhart, E., *Polynómes arithmétiques et Méthode des Polyèdres en Combinatoire.* International Series of Numerical Mathematics, vol. 35, Birkhäuser Verlag, Basel/Stuttgart, 1977.

8. Fisher, D. C., Collins, K. L., and Krompart, L. B., Problem 10406 and solution, *American Mathematical Monthly* **104** (1997), 572–573.
9. Funkenbusch, W. W., From Euler's formula to Pick's formula using an edge theorem, *American Mathematical Monthly* **81** (1974), 647–648.
10. Gaskell, R. W., Klamkin, M. S., and Watson, P., Triangulations and Pick's theorem, *Mathematics Magazine* **49** (1976), 35–37.
11. Grünbaum, B., and Shephard, G. C., Pick's theorem, *American Mathematical Monthly* **100** (1993), 150–161.
12. Iseri, H., An exploration of Pick's theorem in space, *Mathematics Magazine* **81** (2008), 106–115.
13. Liu, A. C. F., Lattice points and Pick's theorem, *Mathematics Magazine* **52** (1979), 232–235.
14. Macdonald, I. G., The volume of a lattice polyhedron, *Proceedings of the Cambridge Philosophical Society* **59** (1963), 719–726.
15. Nyblom, M. A., Pick's theorem and greatest common divisors, *Mathematical Spectrum* **38** (2005/06), 9–11.
16. Pick, G., Geometrisches zur Zahlenlehre, *Sitzungsberichte Lotos (Prag) Naturwissenschaftlich-Medizinschen Vereines für Böhmen* **19** (1899), 311–319.
17. Reeve, J. E., On the volume of lattice polyhedra, *Proceedings of the London Mathematical Society* (3rd Ser.), **7** (1957), 378–395.
18. Rosenholtz, I., Calculating surface areas from a blueprint, *Mathematics Magazine* **52** (1979), 252–256.
19. Stewart, Ian, *Another Fine Math You've Got Me Into ....* Dover, New York, 2003.
20. Varberg, D. E., Pick's theorem revisited, *American Mathematical Monthly* **92** (1985), 584–587.
21. Weaver, C. S., Geoboard triangles with one interior point, *Mathematics Magazine* **50** (1977), 92–94.

2.11 Problems

1. Explain why a line in the plane passes through zero, one, or infinitely many lattice points. Give an example of each case.

2. (a) Show that the number of lattice points on the segment joining the lattice points (a_1, b_1) and (a_2, b_2) is

$$\gcd(a_2 - a_1, b_2 - b_1) + 1.$$

(b) Show that the number of boundary lattice points for the lattice triangle with vertices $(0,0)$, (a_1, b_1), and (a_2, b_2) is

$$\gcd(a_1, b_1) + \gcd(a_2, b_2) + \gcd(a_2 - a_1, b_2 - b_1).$$

3. Heron's formula asserts that the area A of a triangle with sides a, b, and c and semiperimeter $s = (a + b + c)/2$ is

$$A = \sqrt{s(s-a)(s-b)(s-c)}.$$

Verify that the central triangle in Figure 2.3 has sides $\sqrt{37}$, $\sqrt{41}$, and $\sqrt{10}$. Then apply Heron's formula to show that the area of the triangle is $A_2 = 19/2$.

4. (a) Exhibit a 5-by-5 lattice square with 36 lattice points.

   (b) Prove that a 5-by 5 lattice square has at most 36 lattice points. Hint: How many lattice points could occur on an edge of the square?

   (c) Exhibit a 5-by-5 lattice square with 28 lattice points.

   (d) Prove that a 5-by 5 lattice square has at least 28 lattice points.

5. (a) Explain why no two-dimensional lattice triangle is equilateral. Hint: The area of an equilateral triangle with edge length a is $(\sqrt{3}/4)\, a^2$.

   (b) Is there an equilateral three-dimensional lattice triangle?

   (c) Is there a two-dimensional regular lattice hexagon?

   (d) Is there a three-dimensional regular lattice hexagon?

6. How many ways are there to make change for a dollar from a supply of quarters, dimes, nickels, and pennies if at least one coin of each denomination must be used?

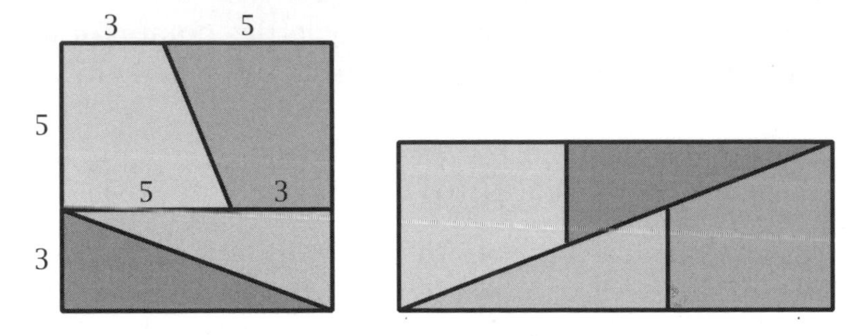

Figure 2.21: A dissection puzzle

7. Is 64 equal to 65? This problem deals with a famous dissection puzzle in which an 8-by-8 square and a 13-by-5 rectangle seem to be partitioned into the same four pieces (Figure 2.21).

   (a) What does Problem 2 say about the number of lattice points on the diagonal from the lower left vertex $(0, 0)$ to the upper right vertex $(13, 5)$ in the rectangle?

   (b) What is the area of the lattice triangle with vertices $(0, 0)$, $(8, 3)$, and $(13, 5)$?

   (c) Use a large drawing of the partitioned rectangle to account for the extra 1 unit of area.

   (d) Create similar dissection puzzles for 13-by-13 and 21-by-21 squares.

8. Let L_N be the lattice point enumerator for a lattice polygon P in two dimensions. We saw that L_{-1} counts the number of lattice points in the interior of P. Show that L_{-2} counts the number of lattice points in the interior of P_2.

9. Show that there are 2063 ways to achieve the batting average .313 with at most 2000 at-bats.

Figure 2.22: A lattice path through all lattice points of a rectangle

10. Figure 2.22 shows a 5-by-4 rectangle containing a lattice path from the northwest corner to the southeast corner that visits each of the 30 lattice points without crossing itself. The shaded regions open to the north and east, while the unshaded regions open to the south and west.

    (a) Show that half of the 5-by-4 rectangle is shaded.

    (b) State and prove a generalization for a-by-b rectangles. Hint: Think outside the box!

11. Explain how to apply Pick's formula to show that the area of the triangle with vertices $(0,0)$, $(5/6, 2/3)$, and $(1, 1/6)$ is $19/72$. Hint: N-large your mind.

12. Which angles can occur in a two-dimensional lattice triangle? This problem gives a complete answer. Suppose that $0 < \alpha < 180°$.

    (a) Show that if $\alpha = 90°$ or $\tan(\alpha)$ is rational, then there is a lattice triangle with angle α. (Recall that a rational number is a ratio of two integers.)

    (b) Show that if angle α occurs in a two-dimensional lattice triangle, then either $\alpha = 90°$ or $\tan(\alpha)$ is a rational number. Hint: The trigonometric identity for the tangent of the difference of two angles is

$$\tan(\theta_2 - \theta_1) = \frac{\tan(\theta_2) - \tan(\theta_1)}{1 + \tan(\theta_2)\tan(\theta_1)}.$$

13. Figure 2.23 shows a lattice polygon with two holes. The holes are themselves lattice polygons, whose boundaries are disjoint from one another and from the outer boundary.

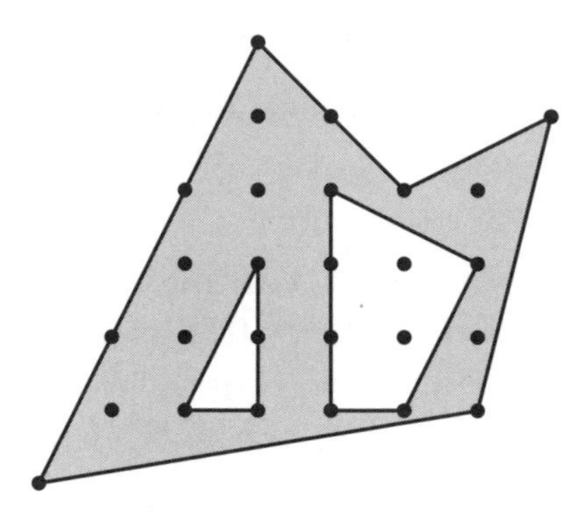

Figure 2.23: A lattice polygon with two holes

(a) Show that the area of the holey polygon is 17.

(b) Show that the area of a holey lattice polygon with L lattice points, B boundary lattice points, and h holes is

$$L - \tfrac{1}{2}B - 1 + h.$$

In Figure 2.23, we have $L = 25$, $B = 18$, and $h = 2$. Note that the lattice points on the boundaries of the holes contribute to B.

14. Show that the right triangular pyramid with vertices

$$(0,0,0),\ (1,0,0),\ (0,1,0),\quad \text{and}\quad (0,0,1)$$

has lattice point enumerator

$$L_N = \frac{(N+1)(N+2)(N+3)}{6}.$$

15. Show that the square pyramid with vertices

$$(0,0,0),\ (1,0,0),\ (0,1,0),\ (1,1,0),\quad \text{and} \quad (0,0,1)$$

has lattice point enumerator

$$L_N = \frac{(N+1)(N+2)(2N+3)}{6}.$$

16. A convex lattice polyhedron P has L lattice points, I interior lattice points, and volume V. Show that the lattice point enumerator of P is

$$L_N = VN^3 + \left(\frac{L-I}{2} - 1\right)N^2 + \left(\frac{L+I}{2} - V\right)N + 1.$$

17. The *triangular lattice* is built from equilateral triangles, each with area 1, as shown in Figure 2.24.

    (a) Show that the area of the lattice polygon in the figure is 11.

    (b) Suppose that a lattice polygon has L lattice points and B boundary lattice points. Show that

$$\text{area of polygon} = 2L - B - 2.$$

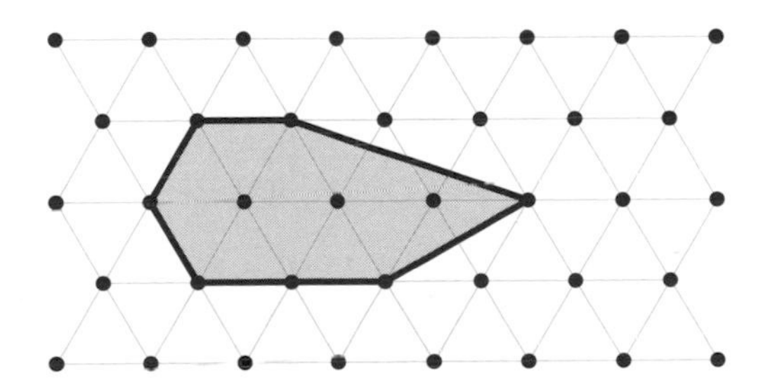

Figure 2.24: A lattice polygon in the triangular lattice

18. A triangle whose side lengths are all integers is an *integer triangle.* Represent each integer triangle by the triple of side lengths $(x, y, p - x - y)$, where p is the perimeter and $x \geq y \geq p - x - y$. For instance, the three integer triangles with perimeter 12 are $(4, 4, 4)$, $(5, 4, 3)$, and $(5, 5, 2)$.

    (a) List the 12 integer triangles with perimeter 24.

    (b) What is the relationship between the triangles in part (a) and the 12 dark lattice points in Figure 2.25?

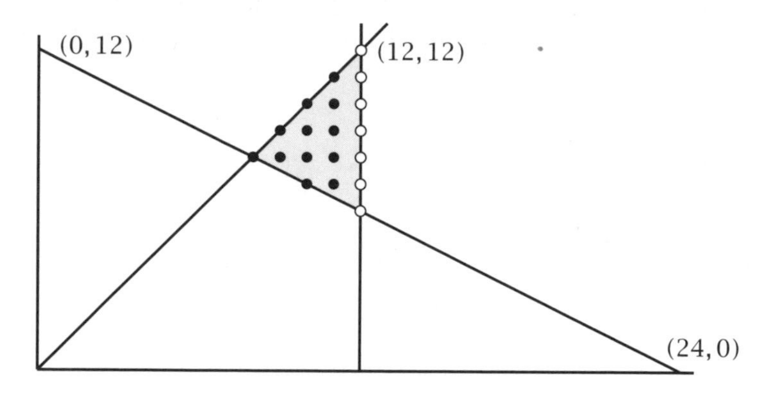

Figure 2.25: Lattice points for integer triangles with perimeter 24

    (c) How many integer triangles have perimeter 36?

    (d) Show that if p is divisible by 12, then there are $p^2/48$ integer triangles with perimeter p.

Change is good, but dollars are better.
ANONYMOUS

When we try to pick out anything by itself,
we find it hitched to everything else in the universe.
JOHN MUIR

3

How to Guard an Art Gallery

I found I could say things with color and shapes
that I couldn't say any other way—
things I had no words for.
GEORGIA O'KEEFFE

3.1 The Sunflower Art Gallery

Figure 3.1 shows the unusual floor plan of the Sunflower Art Gallery and the locations of four guards. Each guard is stationary but can rotate in place to scan the surroundings in all directions. Guards cannot see through walls or around corners. Every point in the gallery is visible to at least one guard, and theft of the artwork is prevented. Of course, it

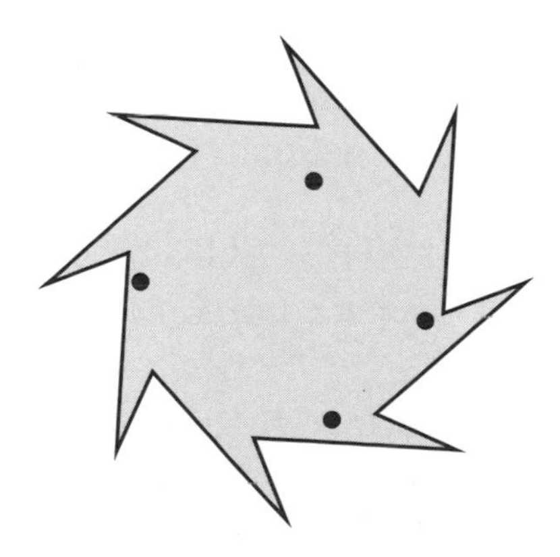

Figure 3.1: The Sunflower Art Gallery

would be more economical to protect the gallery with fewer guards, if possible.

Question. What is the smallest number of guards required to protect the Sunflower Art Gallery?

Removing any one of the four guards in Figure 3.1 leaves part of the gallery unprotected. Nonetheless, it *is* possible to protect the gallery with three suitably positioned guards. We can dismiss the lowest guard if we move the leftmost guard slightly downward. However, we cannot get by with two guards. To see why, consider the eight outer corners of the gallery. It is not possible for one guard anywhere in the gallery to keep an eye on more than three of these corners. So two guards could protect at most six of the eight outer corners.

The Sunflower Art Gallery has 16 walls and needs three guards. This raises a broader question.

Question. What is the smallest number of guards needed to protect any 16-walled gallery, regardless of its shape?

Our art gallery questions involve issues in *computational geometry,* a large and active field that blends geometry with ideas from discrete mathematics and optimization. Applications of computational geometry include:

- Barcode scanners that read prices at grocery stores
- Digital special effects common in today's movies and video games[1]
- Calculations performed by global positioning satellite (GPS) receivers to determine location, speed, and direction

[1]The special case of representing straight lines on a computer monitor is the topic of Chapter 4.

- Algorithms executed by machines and robotic arms on assembly lines to carry out complex tasks in a specific order
- Computerized fingerprint recognition schemes used in security systems and forensics

At the core of most problems in computational geometry is a connection between theory and algorithms. The theory describes or defines desired geometric configurations, which the algorithms construct using known mathematical procedures.

Theoretical and algorithmic issues are tightly linked in art gallery problems. For instance, we will discover a theorem asserting that every w-walled art gallery can be protected by at most $w/3$ guards. Our demonstration of this result leads to an algorithm telling us exactly where to post the guards. We also look at several variations, including an unsolved three-dimensional guarding problem.

3.2 Art Gallery Problems

Let us define our terms carefully. For our purposes, an *art gallery* is a polygon in the plane.[2] The polygon need not serve as the floor plan of any real-world art gallery. An art gallery includes the interior region as well as the boundary segments—the *walls*. We let G denote an arbitrary art gallery and write G_w for an art gallery with w walls.

Let p be any point in an art gallery. The point q is *visible* to p provided the line segment joining p and q does not exit the gallery. (We also assume that every point is visible to itself.) The segment represents the sight line of a guard. A set of guards protects an art gallery provided every point in

[2]More precisely, an art gallery is a *simple polygon*. We exclude polygons with holes, boundaries that cross, and other oddities.

the gallery is visible to at least one guard. Note that a guard at a corner protects the two adjacent walls.

Example 1. (a) The four guards in Figure 3.1 protect the Sunflower Art Gallery.

(b) The Sunflower Art Gallery is not protected by guards at the eight outer corners (Figure 3.2). Even though all of the walls are protected, a region in the center of the gallery remains invisible to all the guards.

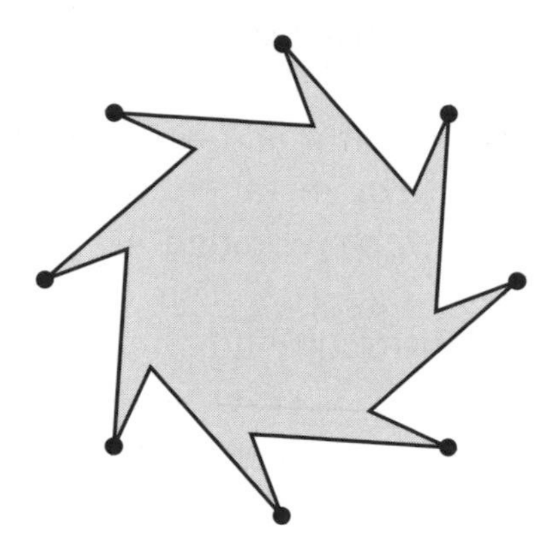

Figure 3.2: The eight guards protect the walls, but not the interior

(c) Each gallery in Figure 3.3 is protected by one or two guards, as shown.

An art gallery is *convex* provided every point in it is visible to every other point. A convex gallery is easy to guard; a guard can be posted anywhere in the gallery. Every triangle is convex, as are the first two galleries in the top row of Figure 3.3. The other galleries in the figure are nonconvex.

Galleries in Particular

Our desire to post as few guards as possible raises two general problems about art galleries. The first problem deals

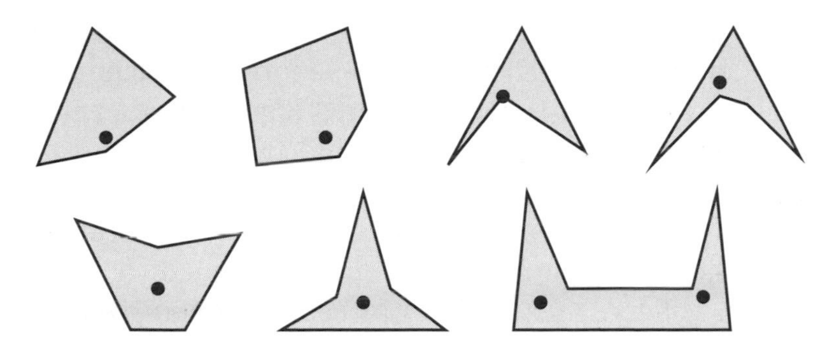

Figure 3.3: The first two galleries in the top row are convex

with specific galleries, and the second deals with all galleries with a fixed number of walls. These are generalizations of the two questions we posed earlier. Let

$$\text{guard}(G) = \text{the minimum number of guards needed to protect the art gallery } G.$$

Gallery problem 1. Find the value of guard(G) for every art gallery G. In other words, find the minimum number of guards needed to protect every art gallery.

Example 2. (a) A convex gallery G satisfies guard$(G) = 1$.

(b) We have seen that the Sunflower Art Gallery G_{16} satisfies guard$(G_{16}) = 3$.

To show that guard$(G) = g$, we must demonstrate two facts:

- The gallery G can be protected by g guards.
- The gallery G cannot be protected by fewer than g guards.

The first fact implies that guard$(G) \leq g$, while the second gives guard$(G) \geq g$. The second fact becomes increasingly difficult to demonstrate as the number of walls increases and the shape of the gallery becomes more complicated.

Ideally, we would have an efficient algorithm that takes an arbitrary gallery G as its input and produces the value of guard(G) as its output. Such an algorithm could be carried out by a computer (or a patient, careful person) to determine the minimum number of guards needed to protect any given gallery. Researchers in *computational complexity,* an advanced area of discrete mathematics, have strong evidence that we will never find an efficient algorithm of the desired type. The crux of the matter is that the number of essentially different guard configurations to examine increases exponentially as a function of the number of walls. Any proposed general algorithm becomes effectively worthless, even with the fastest computers available. In this sense, gallery problem 1 remains unsolved.

Galleries in General

Now suppose we know an art gallery has w walls, but we do not know its exact shape. Let

$$g(w) = \text{the maximum number of guards required among all art galleries with } w \text{ walls.}$$

In other words, $g(w)$ is the maximum value of guard(G_w) among all w-walled galleries G_w.

Example 3. (a) Any triangular art gallery can be protected with one guard. Therefore, $g(3) = 1$.

(b) The Sunflower Art Gallery has 16 walls and requires three guards. Therefore, $g(16) \geq 3$. We cannot conclude that $g(16) = 3$ since there could be a 16-walled gallery that requires more than three guards. In fact, we will soon see a 16-walled gallery requiring five guards.

Gallery problem 2. Find the value of the function $g(w)$ for $w = 3, 4, 5, \ldots$. In other words, find the largest number of guards required among all w-walled art galleries.

To show that $g(w) = g$, we must demonstrate two facts:

- Every w-walled gallery can be protected by g guards.
- There is a w-walled gallery that cannot be protected by fewer than g guards.

The first fact shows that $g(w) \leq g$, while the second shows that $g(w) \geq g$. We will solve gallery problem 2 by establishing both facts. Naturally, we must first find the correct relationship between g and w.

Crown Galleries

To establish a lower bound for $g(w)$, we construct "hard to guard" galleries—those that require at least as many guards as any other gallery with the same number of walls.

We have already noted that $g(3) = 1$. Also, $g(4) = 1$ since a convex quadrilateral clearly requires just one guard, and a nonconvex quadrilateral can be protected by posting one guard at the corner with the largest interior angle (see Figure 3.3). Moreover, it is not difficult to convince oneself that $g(5) = 1$.

The situation is more complicated for galleries with at least six walls, but we can take a hint from the nonconvex, "horned" hexagonal art gallery in Figure 3.3. Because no lone guard can possibly cover both of the two upper corners, we know that $g(6) \geq 2$. The crown-shaped galleries in Figure 3.4 extend this idea. The Crown Gallery G_{3t} has t tines[3] and $3t$ walls and requires at least t guards since

[3]The crown with one tine is more suitable for a dunce than a prince.

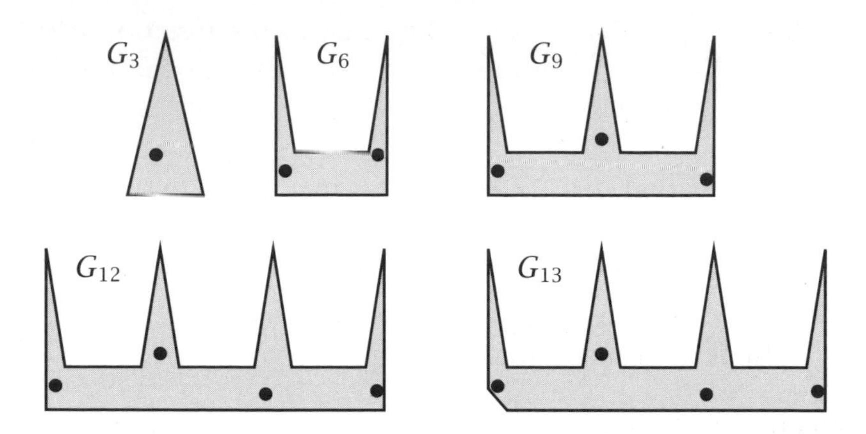

Figure 3.4: The Crown Gallery G_w requires $\lfloor w/3 \rfloor$ guards

no guard can see more than one of the uppermost corners. Therefore, guard$(G_{3t}) \geq t$ and

$$g(3t) \geq t.$$

If w is one more than a multiple of 3, say, $w = 3t + 1$, then we put a small dent in the crown G_{3t} to produce the gallery G_{3t+1}. Figure 3.4 includes the dented gallery G_{13}, for instance. Because G_{3t+1} requires t guards, we have

$$g(3t+1) \geq t.$$

Similarly, if $w = 3t + 2$, a twice-dented crown shows that

$$g(3t+2) \geq t.$$

Some notation helps us state our findings concisely. The *floor* of the real number x is

$$\lfloor x \rfloor = \text{the largest integer less than or equal to } x.$$

So the floor function "rounds down." For example, $\lfloor 16/3 \rfloor = \lfloor 17/3 \rfloor = 5$ and

$$\left\lfloor \frac{3t}{3} \right\rfloor = \left\lfloor \frac{3t+1}{3} \right\rfloor = \left\lfloor \frac{3t+2}{3} \right\rfloor = t$$

for each positive integer t. Our crown-shaped galleries thus give the lower bound

$$g(w) \geq \left\lfloor \frac{w}{3} \right\rfloor.$$

3.3 The Art Gallery Theorem

We are now ready to solve the second gallery problem. The answer confirms that the crown-shaped galleries are indeed the hardest to guard.

Art gallery theorem. We have

$$g(w) = \left\lfloor \frac{w}{3} \right\rfloor \qquad \text{for } w = 3, 4, 5, \ldots.$$

In other words, $\lfloor w/3 \rfloor$ guards are sufficient and sometimes necessary to protect an art gallery with w walls.

The art gallery theorem was first stated and proved by Vasek Chvátal in 1975 in response to a query from Victor Klee (1925–2007), an expert in combinatorial problems with a geometric flavor. We have already discovered that $g(w) \geq \lfloor w/3 \rfloor$. Chvátal's crucial contribution was to establish the reverse inequality by showing that every w-walled gallery can be protected by at most $\lfloor w/3 \rfloor$ guards. His proof uses mathematical induction on the number of walls (the validity of the inequality for galleries with w walls is deduced from its validity for galleries with fewer walls) and requires some care in its execution. Problem 24 at the end of this chapter outlines his argument.

A Colorful Idea

Steve Fisk produced a new and colorful proof of the art gallery theorem in 1978. His ingenious argument is less sophisticated than Chvátal's and has a visual appeal. He assigns

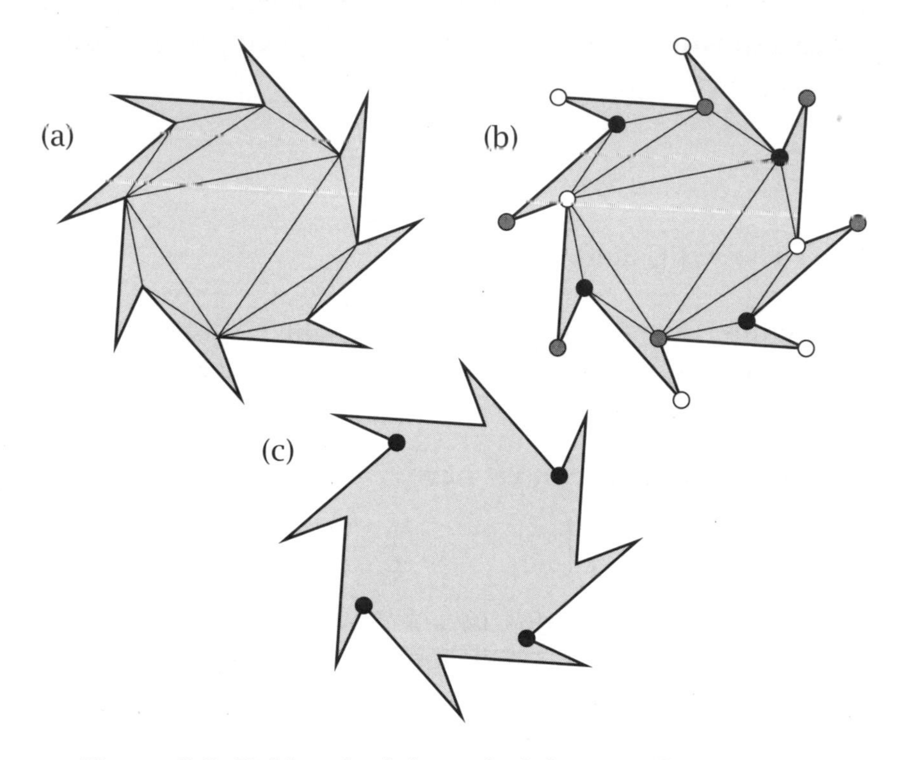

Figure 3.5: Fisk's colorful proof of the art gallery theorem

colors to the corners of the art gallery in a special way and then posts guards based on the arrangement of colors. Figure 3.5 illustrates the steps for the Sunflower Art Gallery.

First, partition the gallery into triangles by inserting suitable noncrossing diagonals, as in (a). The diagonals *triangulate* the gallery.[4] Then assign one of three colors—black, gray, or white, say—to each of the w corners so that every triangle has one corner of each color. The resulting configuration is called a *polychromatic 3-coloring* of the triangulation. In (b) we have $w = 16$, and there are four black, six

[4]The vertices of the triangles must be corners of the gallery; interior vertices are forbidden. More general triangulations appear in Chapter 2 and in Problem 18 of Chapter 1.

gray, and six white corners. Finally, if we post guards at the four black corners, then every triangle is certainly protected (since every triangle has a black corner), and hence the entire gallery is protected by the guards in (c). The six white corners or the six gray corners also protect the gallery, but the black corners give us fewer guards in this case.

The same argument applies to any w-walled gallery. In a polychromatic 3-coloring of a triangulation, the least frequently used color occurs at most $\lfloor w/3 \rfloor$ times. Guards at those corners protect every triangle and hence the entire gallery.

3.4 Colorful Consequences

Fisk's colorful proof of the art gallery theorem has several consequences.

How to Guard an Art Gallery: An Algorithm

The colorful proof not only guarantees that $\lfloor w/3 \rfloor$ guards suffice to protect any w-walled gallery but also tells us exactly where to post at most $\lfloor w/3 \rfloor$ guards. Briefly, triangulate, color, and post. The *art gallery algorithm* (Algorithm 3.1) formalizes the process.

The interactive site

`cut-the-knot.org/Curriculum/Combinatorics/Chvatal.shtml`

lets you build your own art galleries and triangulations; the applet then produces a 3-coloring and posts the guards.

What Is an Algorithm?

We have seen the first of several algorithms in this book, and it is appropriate to make a few comments here. An algorithm is a recipe—a list of precise instructions—that be-

---

Algorithm 3.1. Art gallery algorithm

Input: art gallery G_w with w walls
Output: positions for at most $w/3$ guards that protect G_w

1. Triangulate G_w by inserting suitable diagonals.
2. Find a polychromatic 3-coloring of the corners of the triangulation.
3. Post guards at the corners with the least frequently used color.

---

gins with given ingredients (the input) and ends at a specified goal (the output). Algorithms occur throughout mathematics but are especially prevalent in discrete mathematics. Reading and writing a well-constructed algorithm hones our problem-solving skills and focuses our attention on the essential aspects of a mathematical problem.

An algorithm to be used in a real-world application must be written with great formality in a suitable programming language to avoid the glitches for which computers have become infamous. The algorithms we present in this book are intended for human edification, not actual computer implementation. They are therefore less formal and written in ordinary English.

Nice Try, But ...

Now that we know that $\lfloor w/3 \rfloor$ guards suffice to protect any w-walled gallery, it is natural to seek a simpler and direct process to post the guards. For instance, one attempt to avoid the fuss of triangulation and coloring in Algorithm 3.1 merely posts guards at every third corner of the gallery.

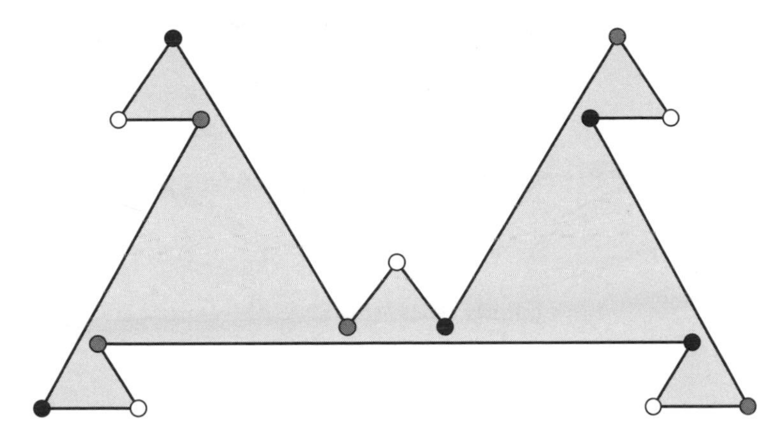

Figure 3.6: Guards at every third corner fail to protect the gallery

This naive strategy works for many galleries but fails for others. Consider the 15-walled gallery in Figure 3.6 with successive corners colored in a repeating black-white-gray pattern. If we post guards at all the black corners, then part of the gallery is unprotected. Guards at the white or gray corners also fail to protect the entire gallery.

Cornered Guards

The art gallery algorithm does not necessarily post the minimum number of guards needed to protect a given gallery. For instance, the algorithm posts four guards in the Sunflower Art Gallery in Figure 3.5, and we know that three guards suffice. In other words, Algorithm 3.1 solves the second gallery problem but not the first.

The algorithm also shows that we can protect the gallery G_w by placing at most $\lfloor w/3 \rfloor$ guards at *corners.* There is no need to place guards in the interior of the gallery—although such placements might be helpful in trying to find the true minimum number of guards required to protect G_w. For instance, one suitably placed guard along the "horizon" of

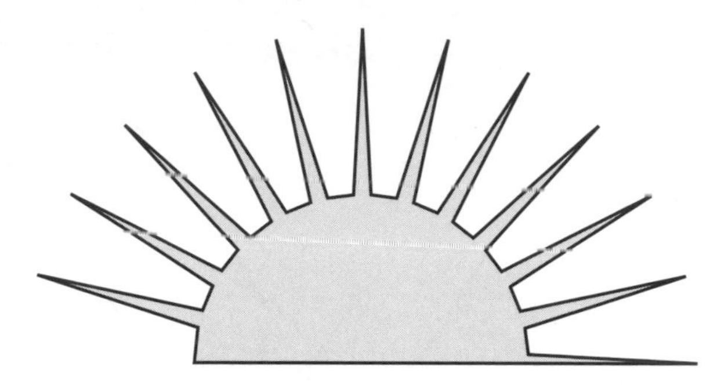

Figure 3.7: The Sunrise Gallery

the Sunrise Gallery in Figure 3.7 protects the entire gallery. But if we must place the guards at corners, then many more guards are needed to cover each ray of the sun.

3.5 Triangular and Chromatic Assumptions

A careful reader might object that our colorful proof of the art gallery theorem is incomplete because we relied on two assumptions without justifying them. Both assumptions are so plausible, you likely did not identify them as potential causes for concern.

Assumption 1. Every art gallery has a triangulation.

Assumption 2. Every triangulation has a polychromatic 3-coloring.

It is wise to question the assumptions we make in mathematics. Plausible assertions sometimes turn out to be false on closer inspection (as we will see later in this chapter), invalidating an entire line of reasoning. The correctness of the colorful proof art gallery theorem is not in doubt, however. Triangulations and polychromatic 3-colorings occur in several contexts in discrete mathematics and have been studied in detail. Rigorous justifications of both assumptions have been known for a long time. See Problems 22 and 23 for a verification of assumption 1.

Polychromatic 3-*Colorings*

There is a convincing, constructive way to verify assumption 2. To show that the particular triangulation of the gallery G_9 in Figure 3.8 has a polychromatic 3-coloring, we first assign three different colors to the corners of an arbitrary triangle, say, triangle acd, as shown. Since each triangle is to contain one corner of each color, corner b must be the same color as corner d. Also, corner f must be the same color as corner c, and then i must be the same color as d. We continue in this manner and eventually produce the desired polychromatic 3-coloring for the entire triangulation.

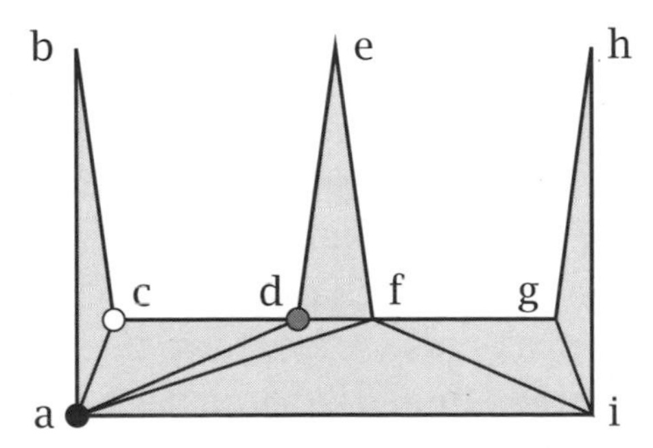

Figure 3.8: The start of a polychromatic 3-coloring

The same process works in general. Once three colors are assigned to the corners of any triangle, the colors for the remaining corners of the gallery are forced.

Degenerate Quadrilaterals

The triangulation of the gallery G_9 in Figure 3.8 illustrates a technical issue that sometimes arises in triangulating a gallery. We maintain that deleting diagonal ad destroys the triangulation even though each remaining region would indeed be triangular. The problem is that the four corners a, c, d, and f of G_9 occur on the boundary of one triangular region. We regard such regions as *degenerate quadrilaterals,* not triangles, and exclude them from our triangulations. This exclusion is necessary for our colorful argument to work.

3.6 Modern Art Galleries

The art gallery theorem has inspired work on related problems in which the rules are changed in some manner to make the model more realistic or more interesting. The changes are of two types. First, we can restrict or relax the allowed shapes for the galleries. Second, we can bestow new powers on the guards or alter their responsibilities. All such variants are referred to as *art gallery problems.* The goals are the same as before. We want to find the minimum number of guards needed and write an efficient algorithm that posts a relatively small number of guards.

The remainder of this chapter is devoted to some of these art gallery problems. A few have been solved, usually by adapting Chvátal's inductive approach or Fisk's colorful argument, but many remain unsolved. The more realistic an art gallery problem, the more difficult it is to discover, to state, and to prove a counterpart to the basic guard formula $g(w) = \lfloor w/3 \rfloor$.

Fortresses, Prisons, and Zoos

Here are some examples of art gallery problems.

In the *fortress problem,* we view a polygon not as an art gallery to be protected against theft from the inside but as a fortress to be alerted to attack from the outside. The goal is to post the minimum number of guards along the fortress walls so that every point outside the fortress is visible to at least one guard.

The *prison yard problem* asks us to post guards on the boundary of a polygon so that every point in the plane—both inside and outside the polygon—is visible to at least one guard. From a mathematical perspective, a prison yard is an art gallery on the inside but a fortress from the outside, a viewpoint presumably not shared by the prison yard's occupants.

In some realistic art gallery problems, the guards are mobile. We can ask for a path of minimum length inside a polygon such that every point in the polygon is visible to some point on the path. Such a path is an efficient route for a lone guard patrolling a large art gallery.

In the *zookeeper problem* we have a collection of disjoint polygons (the animals' cages) inside a large polygon (the zoo). We seek a path of minimum length inside the zoo that meets the boundary of each cage, while avoiding the interior. Such a path traces an efficient and safe route for a zookeeper at feeding time.

The whimsical names bestowed on art gallery problems do not limit the scope of possible applications. For example, the scientists directing the actions of a rover on Mars confront a type of zookeeper problem. The goal is to maneuver the rover to various locations, gather images and measurements of interesting features in the vicinity of the landing spot, and send the data to Earth. There are constraints on time and energy, and steep terrain must be avoided.

3.7 Art Gallery Sketches

We now state some art gallery theorems with proofs omitted.

Galleries with Holes

Most art galleries in the real world contain obstacles that block the sight lines of the guards. We model this situation by allowing *holes* in the interior of the galleries. We assume

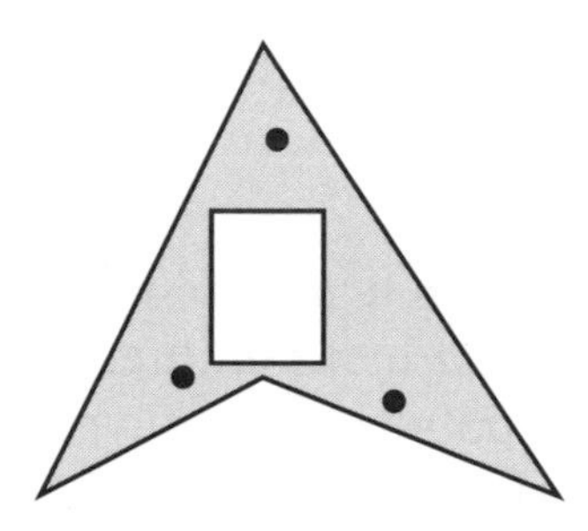

Figure 3.9: A gallery with eight walls and one hole

that each hole is a simple polygon. Guards must not be posted in the interiors of the holes, of course. Figure 3.9 shows a gallery with eight walls and one hole. The gallery is protected by three guards. It is not difficult to verify that the gallery cannot be protected by two guards.

The general problem asks for the minimum number of guards sufficient to protect any gallery with w walls and h holes. Note that the walls surrounding the holes contribute to w. Here is the main theorem in the area.

Theorem. Any art gallery with w walls and h holes can be protected by $\lfloor (w + h) / 3 \rfloor$ guards.

Half-Guards: Restricted Field of Vision

Suppose we want to protect a w-walled gallery with stationary cameras, each of which has a fixed 180° field of vision. We call this type of camera a *half-guard*. It seems likely that more than $w/3$ half-guards might be needed to compensate for the restricted field of vision of the guards. In any case, $2\lfloor w/3 \rfloor$ half-guards suffice since we can use $\lfloor w/3 \rfloor$ pairs of back-to-back half-guards. Surprisingly, we can always get away with just $\lfloor w/3 \rfloor$ half-guards.

Half-guard theorem. Any art gallery with w walls can be protected by $\lfloor w/3 \rfloor$ half-guards.

The half-guard theorem was established by the Hungarian mathematician Csaba Tóth in 2000. The new twist is that suitable corners for the half-guards can no longer be found by a colorful argument. In fact, it is sometimes necessary to place half-guards in the interior or along the walls of the gallery. Figure 3.10 shows a gallery protected by one half-guard along a wall. No corner placement of a lone half-guard does the job for this gallery.

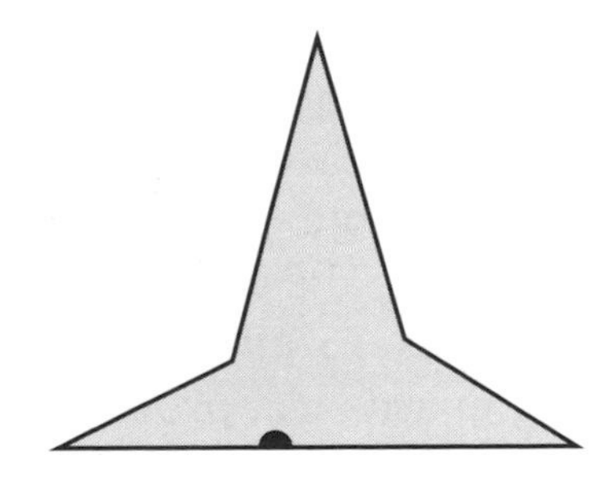

Figure 3.10: A gallery guarded by a half-guard

Rectangulated Galleries: Guarding the Met

Figure 3.11 shows a slightly modified floor plan of one section of the Metropolitan Museum of Art in New York City. (Some walls were adjusted, and a few new doorways were included.) Interior walls partition the large rectangle into rectangular rooms, and each pair of adjacent rooms is joined by

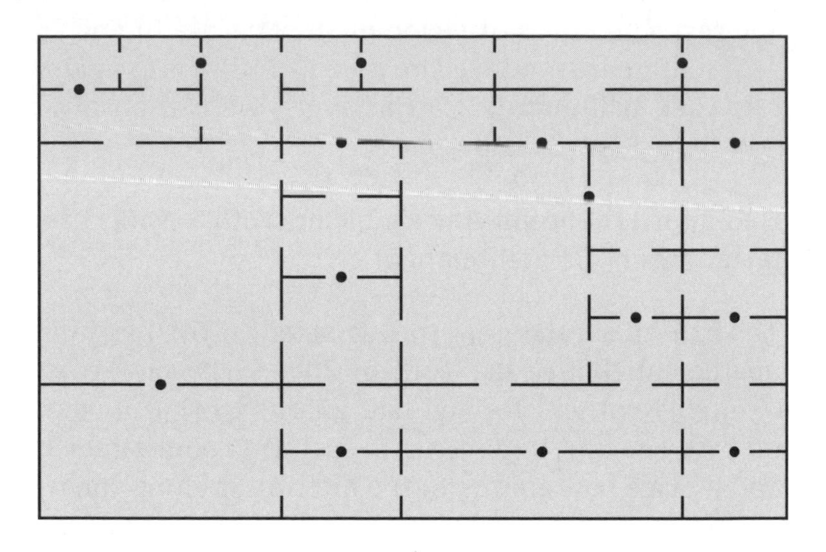

Figure 3.11: A rectangulated gallery with 29 rooms and 15 guards

a narrow doorway. We call this type of configuration a *rectangulated gallery.* As usual, we want to protect the gallery with as few guards as possible. Guards in doorways protect two rooms simultaneously. The gallery in Figure 3.11 has 29 rooms and is protected by 15 guards, which is the best we can hope for since 14 guards can protect at most 28 rooms. In general, if there are r rooms, then at least $\lceil r/2 \rceil$ guards are required. We have used the *ceiling* function, defined by

$$\lceil x \rceil = \text{the smallest integer greater than or equal to } x.$$

It turns out that $\lceil r/2 \rceil$ guards suffice, but this is difficult to prove.

Rectangulated gallery theorem. Any rectangulated gallery with r rooms can be protected by $\lceil r/2 \rceil$ guards, but no fewer.

3.8 Right-Angled Art Galleries

We now examine several art gallery problems in detail and illustrate how art gallery results are discovered and proved.

Adjacent walls in a *right-angled art gallery* meet at right angles, just like the floor plans of most buildings. See Figure 3.12. Each interior angle is 90° or 270°. A right-angled

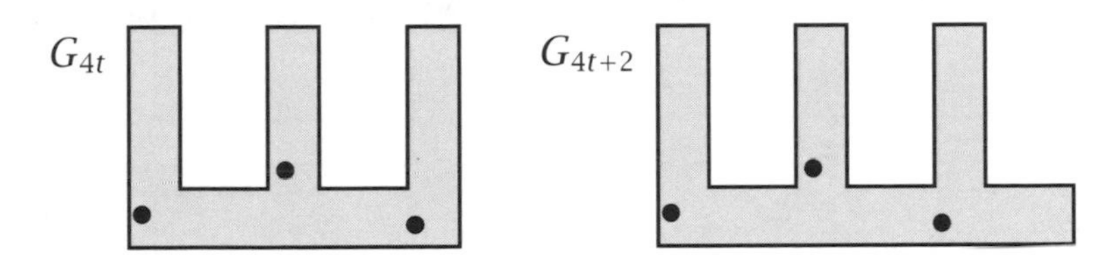

Figure 3.12: The comb-shaped galleries G_{4t} and G_{4t+2}

gallery can be drawn so that the walls run alternately north-south and east-west. It follows that the number of walls must be even.

We let

$g_{\perp}(w)$ = the maximum number of guards required to protect a right-angled art gallery with w walls.

The notation $g_{\perp}(w)$ is pronounced "g perp of w." The subscript is a visual reminder of the perpendicularity of the walls. The comb-shaped galleries in Figure 3.12 play the role of our earlier crown-shaped galleries. Each of the t teeth of the comb adds four more walls and requires one additional guard. It follows that the galleries G_{4t} and G_{4t+2} in the figure require t guards, giving us the lower bound

$$g_{\perp}(w) \geq \left\lfloor \frac{w}{4} \right\rfloor.$$

To establish the reverse inequality, we attempt to modify Fisk's colorful argument. Figure 3.13 depicts a promising strategy. First, partition the right-angled gallery G_w into

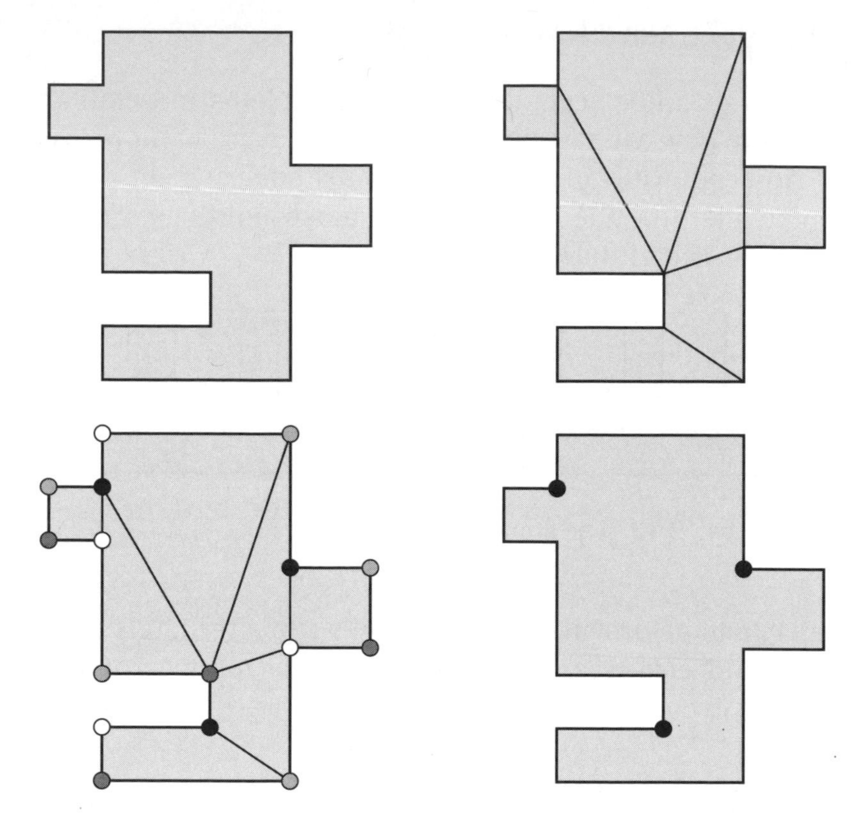

Figure 3.13: A colorful argument for right-angled galleries

quadrilaterals by inserting noncrossing diagonals. The resulting configuration is a *quadrangulation* of G_w. Each quadrilateral has exactly four corners of G_w on its boundary. If three of the corners are collinear, then the quadrilateral has a triangular shape, a degeneracy we now allow. Second, assign one of four colors (black, dark gray, light gray, white) to each of the w corners of G_w so that each quadrilateral has one corner of each color. The least frequently used color in this *polychromatic 4-coloring* occurs at most $\lfloor w/4 \rfloor$ times. Finally, post guards at these corners (the black ones in Figure 3.13); the whole gallery is protected since every quadrilateral has a black corner. It seems we have established the following result.

Right-angled art gallery theorem. We have

$$g_{\perp}(w) = \left\lfloor \frac{w}{4} \right\rfloor \qquad \text{for } w = 4, 6, 8, \ldots.$$

In other words, $\lfloor w/4 \rfloor$ guards are sufficient and sometimes necessary to protect a right-angled art gallery with w walls.

Alas, our colorful argument has a flaw. It fails to post the guards correctly in some situations. Consider the polychromatic 4-coloring of the quadrangulation in Figure 3.14. The three guards at the black corners fail to protect part of the rightmost nook of the gallery. The reason is clear. A guard at a corner of nonconvex quadrilateral might not cover the entire quadrilateral.

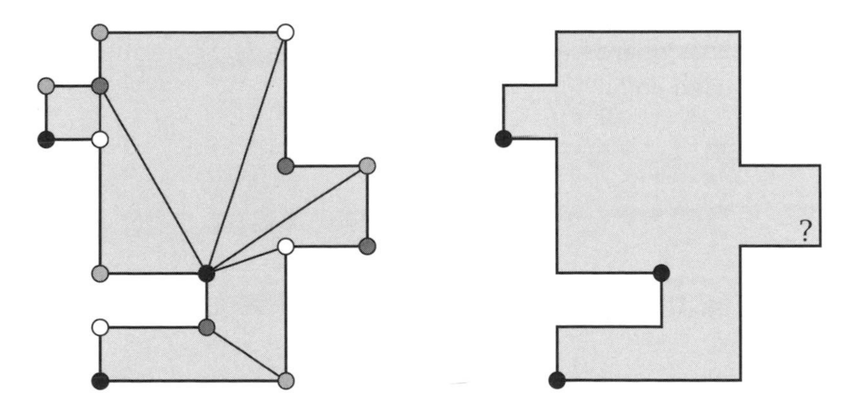

Figure 3.14: A convex quadrangulation is required

To guarantee that the colorful argument works, we must start with a *convex quadrangulation*—one whose quadrilaterals are all convex. Jeffry Kahn, Maria Klawe, and Daniel Kleitman fixed the flaw in the above argument by proving the following result in 1985.

Convex quadrangulation theorem. Every right-angled art gallery has a convex quadrangulation.

The right-angled art gallery algorithm (Algorithm 3.2) formalizes our discussion.

Algorithm 3.2. Algorithm for right-angled art galleries

Input: right-angled art gallery G_w with w walls
Output: positions for at most $w/4$ guards that protect G_w

1. Form a convex quadrangulation of G_w by inserting suitable diagonals.
2. Find a polychromatic 4-coloring of the corners of the quadrangulation.
3. Post guards at the corners with the least frequently used color.

3.9 Guarding the Guards

We now demand that our guards protect one another in addition to the art gallery. Every guard must be visible to at least one other guard. Such configurations protect against an ambush of an isolated guard. We refer to *guarded guards* in this case and study the function

$$gg(w) = \text{the maximum number of guarded guards required to protect an art gallery with } w \text{ walls.}$$

It is not difficult to see that $gg(w) = 2$ for $w = 3, 4, 5, 6$. We must have

$$\left\lfloor \frac{w}{3} \right\rfloor \le gg(w) \le 2\left\lfloor \frac{w}{3} \right\rfloor.$$

The left inequality is clear because we need at least $g(w)$ guards just to cover the gallery. Also, by starting with a feasible configuration of $g(w)$ guards and assigning each guard a nearby partner, we see that no more than $2\,g(w)$ guarded guards are needed.

To determine the formula for $gg(w)$, we might start with another modification to the crown galleries. The New Wave Gallery G_{5t} (shown for $t = 4$ in Figure 3.15) has t waves and $5t$ walls. Note that G_{5t} requires $2t$ guarded guards since each additional wave increases the number of walls by 5 and the number of guarded guards required by 2. We dent the gallery in suitable places if w is not divisible by 5 and conclude that

$$gg(w) \ge \left\lfloor \frac{2w}{5} \right\rfloor \qquad \text{for } w = 5, 6, 7, \ldots.$$

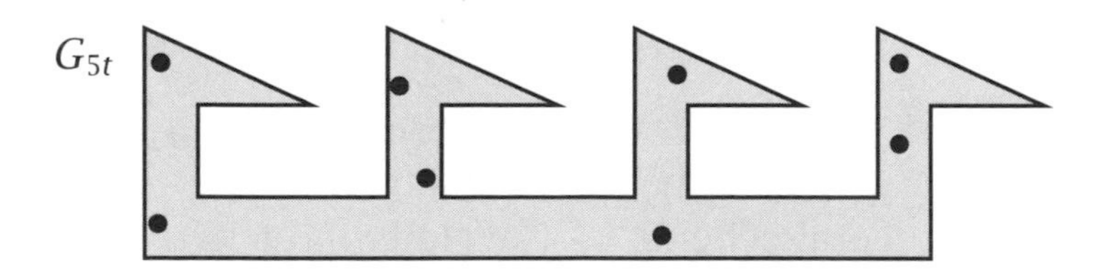

Figure 3.15: The New Wave Gallery G_{5t}

It is tempting to conjecture that $gg(w) = \lfloor 2w/5 \rfloor$, but a rude counterexample intrudes at $w = 12$. The 12-walled gallery in Figure 3.16 requires five guarded guards, contrary to the predicted maximum of $\lfloor 2w/5 \rfloor = \lfloor (2 \times 12)/5 \rfloor = 4$. As

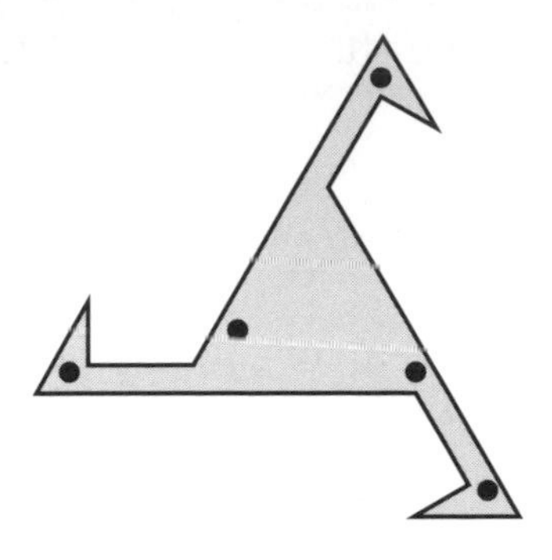

Figure 3.16: A 12-walled gallery requiring five guarded guards

this counterexample suggests, the true formula for guarded guards is more complicated than any we have encountered so far.

Guarded guards theorem. We have

$$gg(w) = \left\lfloor \frac{3w - 1}{7} \right\rfloor \qquad \text{for } w = 5, 6, 7, \ldots.$$

In other words, $\lfloor (3w - 1)/7 \rfloor$ guarded guards are sufficient and sometimes necessary to protect an art gallery with w walls for $w = 5, 6, 7, \ldots$.

Table 3.1 gives some values of the function $gg(w)$. The only known proof of the guarded guards theorem follows the same general scheme of Chvátal's inductive argument

Table 3.1: Number of guarded guards for a gallery with w walls

w	5	6	7	8	9	10	11	12	13	14	15	16	17
$gg(w)$	2	2	2	3	3	4	4	5	5	5	6	6	7

for the original art gallery theorem, but two issues complicate the argument. First, the w-walled galleries that require $\lfloor (3w-1)/7 \rfloor$ guarded guards have complex shapes that are difficult to discover and describe.

Challenge 1. Can you find a 17-walled gallery that requires seven guarded guards? Look at Figure 3.15 if you are stumped.

Second, the inductive step is more subtle than the one used by Chvátal. Is there a pleasant, colorful argument? Nobody has found one yet.

Research problem. Find a colorful, Fisk-like argument for the inequality $gg(w) \le \lfloor (3w-1)/7 \rfloor$.

Guarded Guards in Right-Angled Galleries

The guarded guards formula for right-angled galleries turns out to be less complicated. Let

$gg_{\perp}(w)$ = the maximum number of guarded guards required to protect a right-angled art gallery with w walls.

The Square Wave Gallery G_{6t} (shown for $t = 4$ in Figure 3.17) has t waves and $6t$ walls. Note that G_{6t} requires

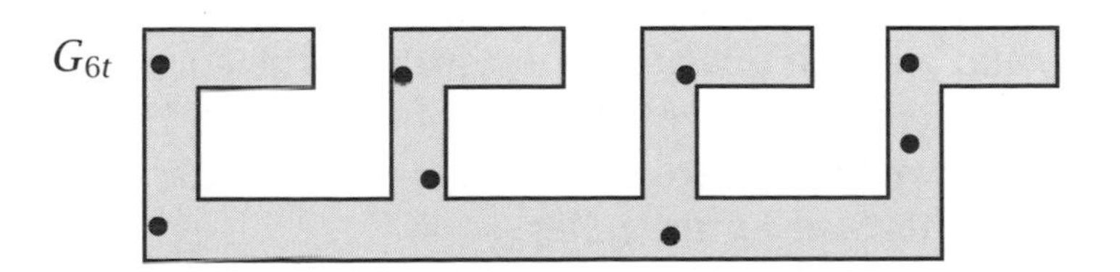

Figure 3.17: The Square Wave Gallery G_{6t}

$2t$ guarded guards since each additional wave increases the number of walls by 6 and the number of guarded guards required by 2. We truncate the gallery in suitable places if w is not divisible by 6 and conclude that

$$gg_{\perp}(w) \geq \left\lfloor \frac{w}{3} \right\rfloor \qquad \text{for } w = 6, 8, 10, \ldots.$$

This time there are no surprises.

Guarded guards theorem for right-angled galleries. We have

$$gg_{\perp}(w) = \left\lfloor \frac{w}{3} \right\rfloor \qquad \text{for } w = 6, 8, 10, \ldots.$$

In other words, $\lfloor w/3 \rfloor$ guarded guards are sufficient and sometimes necessary to protect a right-angled art gallery with w walls for $w = 6, 8, 10, \ldots$.

To show that $\lfloor w/3 \rfloor$ guarded guards are sufficient for a right-angled gallery, we again modify Fisk's colorful argument. Start with a convex quadrangulation of a right-angled gallery G_w, as in Figure 3.18(a). Then triangulate the gallery by inserting a diagonal in each quadrilateral (the thin lines in (b)). The inserted diagonals should "alternate" so that if two quadrilaterals share an edge, then their diagonals do not share a corner. After the diagonal for one quadrilateral is selected arbitrarily, the diagonals for the other quadrilaterals are all forced by this alternating condition. The resulting triangulation has a polychromatic 3-coloring, as shown in (b), and we post guards temporarily at the corners of the least frequently used color—the four black corners in (c). Of course, we have posted at most $\lfloor w/3 \rfloor$ guards, and these guards protect the entire gallery.

Alas, some guards might be invisible to all other guards. The lowest guard in (c) is invisible to the other guards, as is the rightmost guard. We remedy this situation by giving

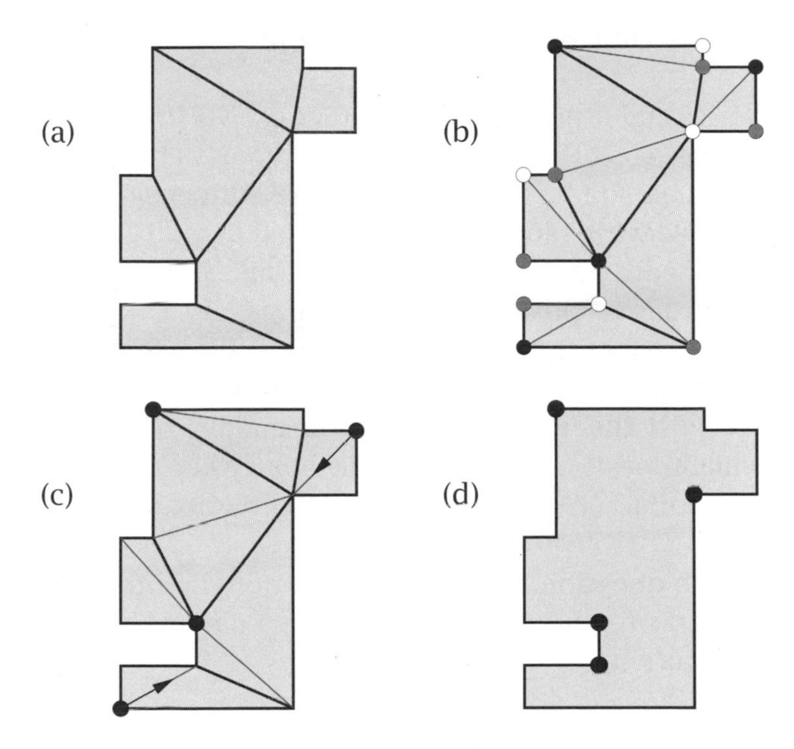

Figure 3.18: Guarded guards for right-angled galleries

marching orders to each guard standing at a corner with exactly one diagonal. Such a guard must march along the diagonal to the opposite corner, as indicated by the arrows in (c). If two or more guards end up at the same corner, then the extra ones are sent home. One can prove that the resulting configuration of guards is indeed guarded and that the entire gallery remains protected, as in (d). We do not give the details.

As with the original art gallery theorem, the resulting configuration of guarded guards need not be minimal. For instance, the process posts four guarded guards in the right-angled gallery in Figure 3.18(d), but three guarded guards suffice; simply dismiss the uppermost guard.

3.10 Three Dimensions and the Octoplex

Art galleries and other buildings in the real world are three-dimensional, a fact that has been conspicuously absent from our discussion so far. Let us model three-dimensional galleries by *polyhedra*—solid shapes bounded by polygons. Familiar polyhedra include cubes, prisms, and pyramids. As usual, we want to post cameras in the gallery so that every point is visible to at least one camera. We use security cameras instead of guards since it will sometimes be necessary to post them on the ceiling, in midair, or at other inconvenient locations. We make the somewhat unrealistic assumption that a camera can see in all directions.

Research question. What is the maximum number of security cameras required to protect a three-dimensional gallery with c corners?

We seek a simple answer to the three-dimensional art gallery problem, similar to our expressions for two-dimensional galleries. However, no one has been able to find such a formula—or even propose a plausible guess—for reasons we will explain soon.

In a two-dimensional gallery the number of corners is equal to the number of walls. But these parameters are unequal for most three-dimensional galleries. (A corner is a point where three or more walls meet.) We would be just as pleased to answer the research question in terms of the number of walls.

The Octoplex

Guards posted at each corner of a two-dimensional gallery certainly protect the whole gallery. Astonishingly, this obvious assertion is false for three-dimensional galleries. This

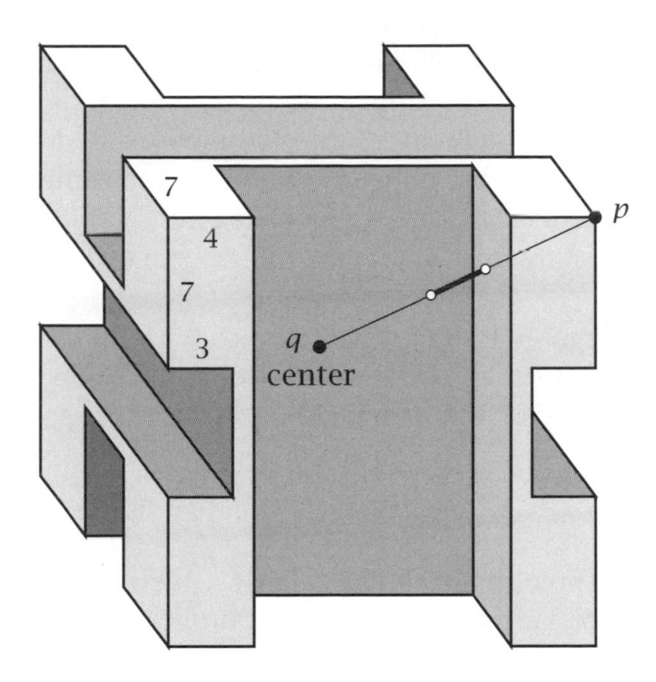

Figure 3.19: The Octoplex

fact frustrates would-be solvers of the three-dimensional problem.

The example in Figure 3.19 is constructed as follows. Start with a 20-by-20-by-20 cube. Remove a rectangular channel 12 units wide and 6 units deep from the center of the front face. There is an identical channel in the back face. The channels in the left and right faces are 6 units wide and 3 units deep, while those in the top and bottom faces are 6 units wide and 6 units deep. The figure that remains is the *Octoplex.* It consists of eight 4-by-7-by-7 theaters connected to one another and to a central lobby by passageways 1 unit wide. The Octoplex has 56 corners and 30 walls.

Claim. Even if we post a camera at every corner, part of the Octoplex is unprotected.

To see why the claim is true, observe that the center point q of the Octoplex is not visible to a camera at the corner p in Figure 3.19 since the direct line of sight from p to q exits and then reenters the Octoplex, as shown. Similar reasoning shows that point q is hidden from cameras at the other 55 corners. In fact, there is a small region in the middle of the lobby that is hidden from every corner camera.

Challenge 2. Protect the Octoplex with 25 cameras. At least one of your cameras will not be at a corner. Can you find a way to use fewer cameras?

The Megaplex

For some three-dimensional galleries, the number of cameras required greatly exceeds the number of corners. The Megaplex is formed by a cubical arrangement of m^3 abutting copies of the Octoplex, as shown for $m = 4$ in Figure 3.20. The interior walls of adjacent theaters are removed so that some theaters in the Megaplex are formed by merging two, four, or eight Octoplex theaters. For the sake of clarity, the walls separating each Octoplex from its neighbors are retained in the figure. There are m^3 lobbies in the Megaplex. Also, The many channels and shafts do not cross one another.

Claim. The Megaplex has $c = 24m^2 + 24m + 8$ corners and requires at least $m^3/8$ cameras to protect.

To see why the claim is true, observe that the Megaplex has eight outer corners, and that each of the $3m(m + 1)$ shafts and channels also contributes eight corners. The total number of corners is therefore

$$c = 8 + 3m(m + 1) \times 8 = 24m^2 + 24m + 8.$$

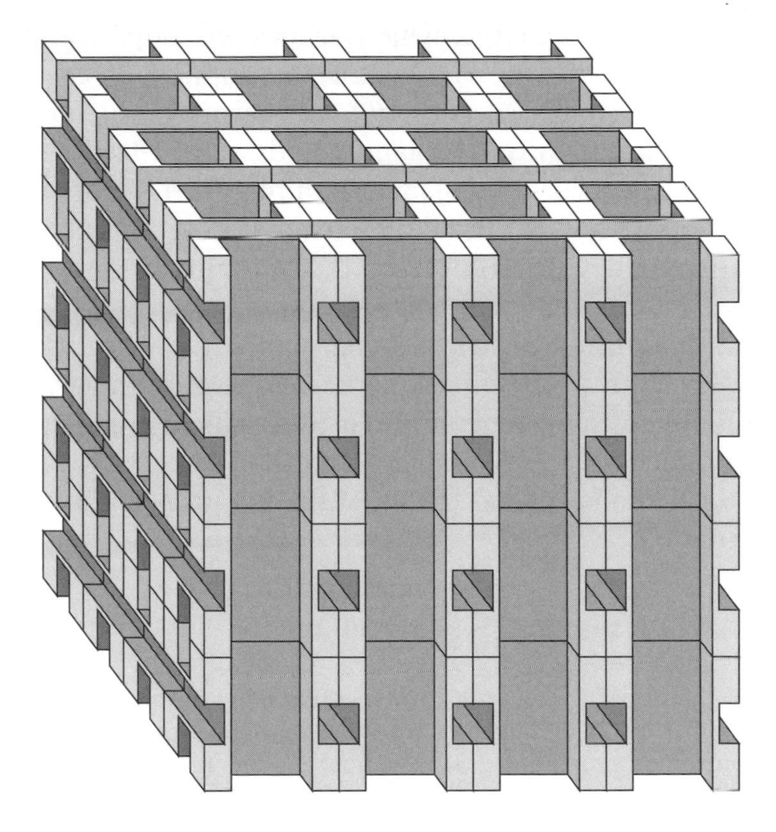

Figure 3.20: The Megaplex

A little experimentation shows that no camera could possibly cover the centers of eight lobbies, and it follows that at least $m^3/8$ cameras are required.

Notice that the ratio of cameras to corners satisfies

$$\frac{\text{cameras}}{\text{corners}} \geq \frac{m^3/8}{24m^2+24m+8} \geq \frac{(m^3-1)/8}{24(m^2+m+1)} \geq \frac{m-1}{192}.$$

The factorization $m^3-1=(m-1)(m^2+m+1)$ was used for the last inequality. If $m \geq 194$, then the number of cameras required exceeds the number of corners of the Megaplex. Moreover, as m increases, the ratio of cameras to corners becomes arbitrarily large, which dashes any hope for a linear upper bound (say, $1000c$) for the number of cameras required. A similar assertion holds in terms of the number of walls. Although our Megaplex with $m > 194$ does not resemble any three-dimensional building in the real world, it does show that the three-dimensional guarding problem is fundamentally different from the two-dimensional problems we have seen.

The horizontal shafts passing completely through the Megaplex from front to back form undesirable "holes." We can eliminate them by erecting a thin wall to close up the back of each shaft. The shafts in other directions can be dealt with similarly, and the essential features of the Megaplex are preserved.

3.11 Notes and References

The original proof of the art gallery theorem is by Chvátal [2]. A more leisurely treatment of Chvátal's proof by mathematical induction appears in Honsberger's book [5]. The book [1] includes a first-hand account of how Fisk discovered his colorful proof [4] on a bus trip in Afghanistan. The right-angled art gallery theorem was first proved by Kahn, Klawe, and Kleitman [6]. Guarded guards are examined in [7]. Żyliński [12] surveys the use of colorful arguments in proving art gallery theorems. The theorem on rectangulated galleries appears in the paper by Czyzowicz et al. [3].

In 1987, O'Rourke wrote *the* book [9] on art gallery theorems, covering both theory and algorithms. Variants examined in O'Rourke's book include mobile guards, the fortress and prison yard problems, and three-dimensional galleries. The encyclopedic tome [11] spans all of computational geometry, and the chapter on art gallery theorems has more than 100 references. Algorithms for art gallery problems (and computational geometry in general) are treated in [10], which also covers computational complexity.

The site

`http://maven.smith.edu/~orourke/TOPP/`

lists unsolved problems in computational geometry.

1. Burger, Edward B., and Starbird, Michael, *The Heart of Mathematics.* Key College Publishing, Emeryville, California, 2000.
2. Chvátal, V., A combinatorial theorem in plane geometry, *Journal of Combinatorial Theory, Series B* **18** (1975), 39–41.
3. Czyzowicz, J., Rivera-Campo, E., Santoro, N., Urrutia, J., and Zaks, J., Tight bounds for the rectangular art gallery problem, *Graph-Theoretic Concepts in Computer Science (Fischbachau 1991),* 105–112, Lecture Notes in Computer Science 570, Springer, Berlin, 1992.
4. Fisk, S., A short proof of Chvátal's watchman theorem, *Journal of Combinatorial Theory, Series B* **24** (1978), 374.
5. Honsberger, Ross, *Mathematical Gems II.* Mathematical Association of America, Washington, DC, 1976.
6. Kahn, J., Klawe, M., and Kleitman, D., Traditional galleries require fewer watchmen, *SIAM Journal of Algebraic and Discrete Methods* **4** (1983), 194–206.
7. Michael, T. S., and Pinciu, V., Art gallery theorems for guarded guards, *Computational Geometry* **26** (2003), 247–258.
8. O'Rourke, J., Galleries need fewer mobile guards: a variation on Chvátal's theorem, *Geometriae Dedicata* **14** (1983), 273–283.
9. O'Rourke, Joseph, *Art Gallery Theorems.* Oxford University Press, Cambridge, U.K., 1987.
10. O'Rourke, Joseph, *Computational Geometry in C,* 2nd ed. Cambridge University Press, Cambridge, U.K., 1998.
11. Sack, J.-R., and Urrutia, J. (eds.), *Handbook of Computational Geometry.* North Holland, Amsterdam, 2000.
12. Żyliński, P., Placing guards in art galleries by graph coloring, *Contemporary Mathematics* **352** (2004), 177–188.

3.12 Problems

The problems deal with two-dimensional art galleries.

1. Find a triangulation of the Sunflower Art Gallery and a polychromatic 3-coloring that leads to a posting of three guards.

2. True or false.

   (a) If G is a convex gallery, then $\text{guard}(G) = 1$.

   (b) If $\text{guard}(G) = 1$, then the gallery G is convex.

(c) If guard$(G) \geq 2$, then G has at least six walls.

(d) If G has at least six walls, then guard$(G) \geq 2$.

3. (a) Exhibit an art gallery with eight walls that has a unique triangulation.

   (b) Exhibit a w-walled gallery that has a unique triangulation for each $w = 3, 4, \ldots$.

4. Which corners are the same color as c when Figure 3.8 is completed to a polychromatic 3-coloring?

5. Let s be the number of $90°$ interior angles in a right-angled gallery with w walls. Show that $w = 2s - 4$. Hint: What is the sum of all the angles in the gallery?

6. Exhibit an art gallery with both of the following properties.

   - The gallery can be protected by one guard.
   - It is possible to post guards at seven corners and not protect the entire gallery.

7. Exhibit an art gallery with both of the following properties.

   - The gallery can be protected by two guards but not by one guard.
   - It is possible to post guards at 29 corners and not protect the entire gallery.

8. Let G_{15} denote the 15-walled gallery in Figure 3.6.

   (a) Protect G_{15} with five guards.

   (b) Show that G_{15} cannot be protected by four guards.

   (c) What is the minimum number of guarded guards needed to protect G_{15}?

9. The galleries in Figure 3.17 show that $gg_{\perp}(w) \geq \lfloor w/3 \rfloor$ when w is divisible by 6. Exhibit right-angled galleries for the cases when $w - 2$ or $w - 4$ is divisible by 6.

10. Find the final configuration of guarded guards if we begin with guards at the four white corners in Figure 3.18(b).

11. Write an algorithm to post guarded guards in right-angled galleries. The input of your algorithm will be a right-angled gallery G_w with w walls ($w \geq 6$), and the output will be the positions of at most $\lfloor w/3 \rfloor$ guarded guards that protect G_w.

12. Post 10 guards in a particular 17-walled art gallery so that the entire gallery is protected, but dismissal of any guard leaves some part of the gallery unprotected.

13. Figure 3.21 shows four guards that protect a rectangulated gallery with six rooms and one rectangular hole.

    (a) Explain why the gallery cannot be protected by three guards.

    (b) Explain why every rectangulated gallery with r rooms and one rectangular hole can be protected by $\lceil (r+1)/2 \rceil$ guards.

    (c) Explain why every rectangulated gallery with r rooms and h rectangular holes can be protected by $\lceil (r+h)/2 \rceil$ guards.

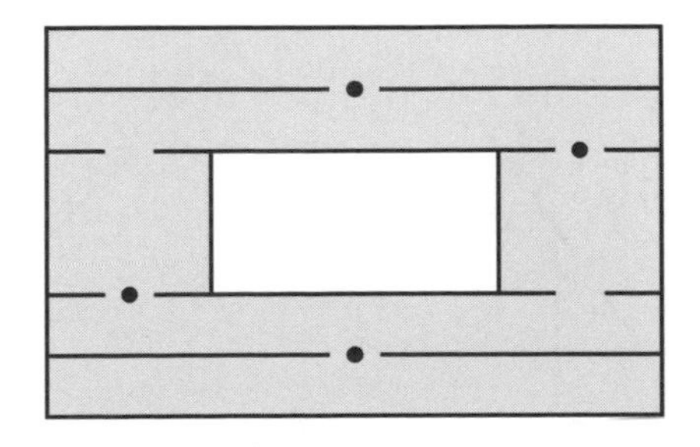

Figure 3.21: A rectangulated gallery with a hole

14. There is an art gallery with $8t$ walls and t holes that requires $3t$ guards for each $t = 1, 2, 3, \ldots$. Figure 3.9 shows such a gallery for $t = 1$. Find a gallery for $t = 2, 3, \ldots$.

15. What is the minimum number of half-guarded needed to protect the Sunflower Art Gallery?

16. Give an example of a gallery that can be protected by one guard but not by one half-guard.

17. Explain why a triangulation of a w-walled gallery must have $w - 2$ triangles and $w - 3$ diagonals.

18. (a) Find a 10-walled gallery requiring four guarded guards.

    (b) Find a 15-walled gallery requiring five guarded guards.

19. The Scorpio Gallery in Figure 3.22 has 17 walls.

    (a) Protect the gallery with seven guarded guards.

    (b) Show that the gallery cannot be protected by six guarded guards.

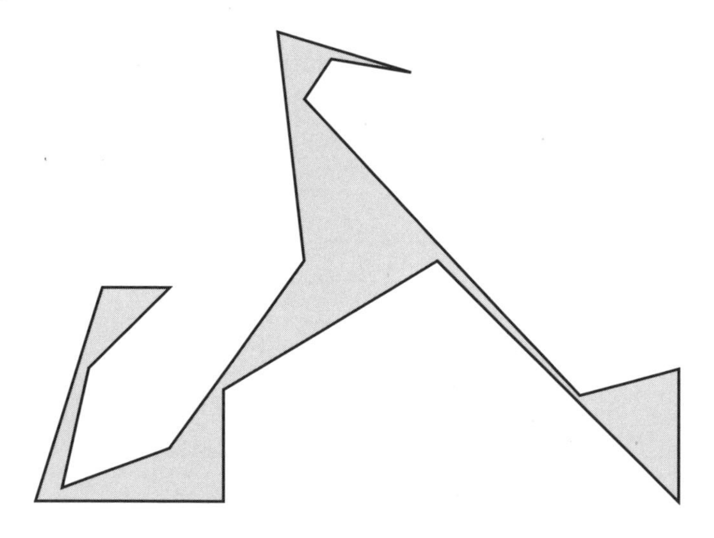

Figure 3.22: The Scorpio Gallery

20. Consider a triangulation of an art gallery with at least four walls. Identify each statement as true or false.

    (a) Every triangle in the triangulation must have at least one side in common with the boundary of the gallery.

    (b) There is a triangle with exactly two sides in common with the boundary.

    (c) There are at least two triangles with exactly two sides in common with the boundary.

21. Consider a triangulation of an art gallery with at least four walls. Let N_k denote the number of triangles with exactly k sides in common with the boundary of the gallery. Of course, $N_k = 0$ for $k \geq 3$.

    (a) Explain why $N_0 + N_1 + N_2 = w - 2$.

    (b) Explain why $N_1 + 2N_2 = w$.

    (c) Show that $3N_0 + 2N_1 + N_2 = 2w - 6$.

    (d) Show that $N_2 = N_0 + 2$.

22. The purpose of this problem is to prove that every art gallery with at least four walls has a diagonal. Let G_w be an art gallery with w walls ($w \geq 4$), and let q be the leftmost corner of the gallery. If more than one corner is leftmost, choose the lowest of these. Let p, q, and r be consecutive corners of the gallery and consider the triangle pqr.

    (a) Suppose that p, q, and r are the only corners of G_w in pqr (including the boundary). Show that segment pr is a diagonal of G_w.

    (b) Suppose that pqr contains at least one other corner of G_w. Show that G_w has a diagonal with one endpoint at q.

23. Show that every art gallery has a triangulation. Hint: Assume that every art gallery with at least four walls has a diagonal (Problem 22).

24. This problem outlines Chvátal's proof of the art gallery theorem. Consider a triangulated art gallery G_w with w walls ($w \geq 4$). Select any diagonal of the triangulation. The diagonal partitions G_w into two galleries with w_1 and w_2 walls.

    (a) Explain why $w_1 + w_2 = w + 2$.

    (b) Show that if $w \geq 6$, then we can always select a diagonal so that $w_1 = 5$, 6, or 7. Hint: Among all diagonals for which $w_1 \geq 5$, choose one for which w_1 is smallest.

    (c) Assume that $w_1 = 5$. Show that

    $$\left\lfloor \frac{w_1}{3} \right\rfloor + \left\lfloor \frac{w_2}{3} \right\rfloor = \left\lfloor \frac{w}{3} \right\rfloor.$$

    Hint: Show that

    $$\left\lfloor \frac{w_2}{3} \right\rfloor = \left\lfloor \frac{w-3}{3} \right\rfloor = \left\lfloor \frac{w}{3} \right\rfloor - 1.$$

    (d) Assume that $w_1 = 5$. Make an inductive hypothesis and apply (c) to show that $g(w) \leq \lfloor w/3 \rfloor$. Chvátal also deals with the more difficult cases $w_1 = 6$ and $w_1 = 7$.

The picture will have charm when each color
is very unlike the one next to it.
LEON BATTISTA ALBERTI

Who will guard the guardians?
JUVENAL

Science is what we understand
well enough to explain to a computer.
Art is everything else we do.
DONALD KNUTH

Mighty is geometry; joined with art, resistless.
EURIPEDES

4

Pixels, Lines, and Leap Years

The shortest distance between two points
is under construction.
LEO AIKMAN

4.1 Pixels and Lines

A computer monitor has a rectangular array of thousands of tiny square cells called *pixels,* each of which is light or dark at any time. In the field of computer graphics, we confront the problem of rendering continuous geometric shapes on the array of discrete pixels in an accurate and pleasing manner. This chapter focuses on the most basic situation, the representation of a straight line—more precisely, a *line segment*—joining two specified points.

Computer line drawing problem. How does a computer construct the best configuration of pixels to represent a line segment joining two specified points?

Figure 4.1 shows a jagged sequence of 14 dark pixels that represent the segment joining the points with Cartesian coordinates $(0,0)$ and $(13,8)$. In our diagrams, the origin is near the lower left corner, the x-axis points rightward, and the y-axis points upward; to reduce clutter, we usually omit the axes. We use unrealistically large pixels to illustrate our ideas more clearly. The center (x, y) of each pixel is a *lattice point;* both coordinates x and y are integers.

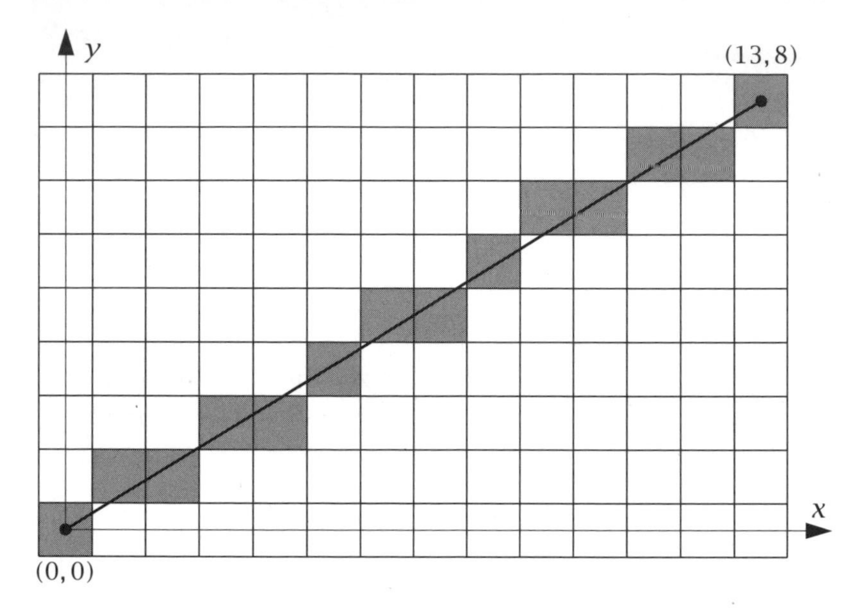

Figure 4.1: Pixels for the segment joining (0, 0) and (13, 8)

The 14 dark pixels in Figure 4.1 are the pixels closest to the segment. The segment has equation $y = (8/13)\,x$, and so the coordinates of the dark pixels are

$$(x, y) = \left(x, \text{ round}\left[\left(\tfrac{8}{13}\right)x\right]\right) \qquad \text{for } x = 0, 1, \ldots, 13.$$

The function round $[Y]$ rounds the number Y to the nearest integer. If Y is halfway between two integers, it is rounded up.

In general, if $0 < v \leq w$, the line segment joining $(0, 0)$ and (w, v) is represented by the pixels

$$(x, y) = \left(x, \text{ round}\left[\left(\tfrac{v}{w}\right)x\right]\right) \qquad \text{for } x = 0, 1, \ldots, w. \quad (4.1)$$

Algorithm 4.1, which is based directly on (4.1), is unsatisfactory for practical use because step 5 relies on operations that slow down the line drawing process. First, computing the expression $(v/w)x$ uses multiplication and division instead of the faster operations of addition and subtraction. Second, the rounding operation occurs outside of the preferred domain of integer arithmetic, where operations can be performed most rapidly.

Algorithm 4.1. A line drawing algorithm

Input: integers v and w with $0 < v \leq w$

Output: $w + 1$ dark pixels that represent the line segment joining $(0, 0)$ and (w, v)

1. Start at pixel $(x, y) = (0, 0)$.
2. Darken pixel (x, y).
3. If $x = w$, then all $w + 1$ pixels are dark. Stop.
4. Replace x by $x + 1$.
5. Let $y = \text{round}\left[\left(\frac{v}{w}\right)x\right]$. Return to step 2.

In the 1960s, Jack Bresenham, a computer scientist at IBM, produced a simple, efficient line drawing algorithm that uses only addition and subtraction of integers and other fast operations to.generate the correct pixels. Bresenham's algorithm is a cornerstone of computer graphics.

In this chapter, we discover the algorithm for ourselves. Although our path is not identical to the one taken by Bresenham, our approach helps us relate the computer line drawing to numerical approximations and the distribution of leap years in calendars. In Chapter 7, we link line drawing to a classic problem in number theory.

4.2 Lines and Distances

We first transform the geometric problem of line drawing to an arithmetic problem involving integers.

A Special Case

Horizontal and vertical segments are easy to represent on a monitor. We will focus on representing a segment joining the origin $(0, 0)$ and a point (w, v) with $0 < v \leq w$. The slope v/w of this segment satisfies

$$0 < \frac{v}{w} \leq 1.$$

Our line drawing algorithm will select one pixel in column x for each $x = 0, 1, \ldots, w$, as in Figure 4.1.

Solving this special case solves the line drawing problem in general. Other segments can be obtained by reflections and translations (parallel shifts), which can be executed quickly and automatically by a computer. For exam-

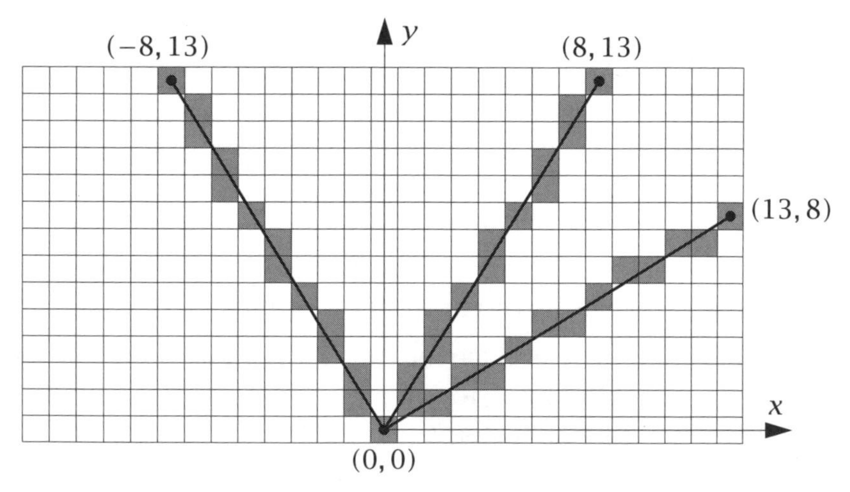

Figure 4.2: Transforming the segment joining $(0, 0)$ and $(13, 8)$

ple, suppose we have already determined the pixels for the segment joining the origin $(0,0)$ and $(13,8)$. Then the segment joining $(20,10)$ and $(33,18)$ is obtained by translating the original segment 20 units horizontally and 10 units vertically. Moreover, interchanging the x- and y-coordinates of the original segment gives the segment joining the origin and $(8,13)$, and then negating the x-coordinates reflects the segment about the y-axis to produce the segment joining the origin and $(-8,13)$, as shown in Figure 4.2.

A Distance Formula

The following formula for the vertical distance from the center of a pixel to a line is the key to our derivation of Bresenham's line drawing algorithm.

Vertical distance formula. The vertical distance from the point (x,y) to the line joining $(0,0)$ and (w,v) is

$$d_{\text{vert}} = \frac{|vx - wy|}{w}.$$

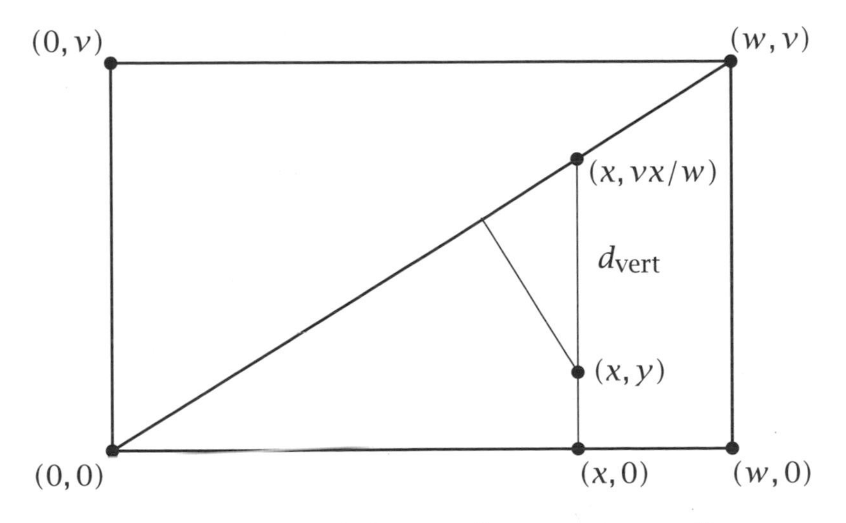

Figure 4.3: Vertical distance from a point to a line

Figure 4.3 shows why the formula is true. The vertical line through (x, y) intersects the slanted line joining $(0,0)$ and (w, v) at the point $(x, vx/w)$. If the point (x, y) falls below the line, then subtraction shows that the vertical distance we seek is

$$d_{\text{vert}} = \frac{vx}{w} - y = \frac{vx - wy}{w},$$

while if (x, y) is above (or on) the line, then

$$d_{\text{vert}} = y - \frac{vx}{w} = -\frac{vx - wy}{w}.$$

The vertical distance formula combines both cases by taking the absolute value.

For the line drawing problem, it is also reasonable to consider the *direct* distance d from (x, y) to the line instead of the vertical distance. However, it is not difficult to show (see Problem 1 for hints) that

$$d = \frac{|vx - wy|}{\sqrt{v^2 + w^2}} = \left(\frac{w}{\sqrt{v^2 + w^2}}\right) d_{\text{vert}},$$

which tells us that within each column the same pixel minimizes both d and d_{vert}.

4.3 Arithmetic Arrays

Suppose we want to select the dark pixels representing the segment joining $(0,0)$ and $(13,8)$. The distance from the pixel with center (x, y) to the segment is

$$d_{\text{vert}} = \frac{|8x - 13y|}{13}.$$

The rectangular array of numbers in Figure 4.4 helps us select the correct pixels. The number $8x - 13y$ occurs in position (x, y) in the array. We know that the pixel (x, y) in

-104	-96	-88	-80	-72	-64	-56	-48	-40	-32	-24	-16	-8	0
-91	-83	-75	-67	-59	-51	-43	-35	-27	-19	-11	-3	5	13
-78	-70	-62	-44	-46	-38	-30	-22	-14	-6	2	10	18	26
-65	-57	-49	-41	-33	-25	-17	-9	-1	7	15	23	31	39
-52	-44	-36	-28	-20	-12	-4	4	12	20	28	36	44	52
-39	-31	-23	-15	-7	1	9	17	25	33	41	49	57	65
-26	-18	-10	-2	6	14	22	30	38	46	54	62	70	78
-13	-5	3	11	19	27	35	43	51	59	67	75	83	91
0	8	16	24	32	40	48	56	64	72	80	88	96	104

Figure 4.4: Pixels in the arithmetic array $A(8, -13)$

column x is closest to the line joining $(0,0)$ and $(13,8)$ exactly when y is chosen to minimize the absolute value of $8x - 13y$. The dark pixel in each column is therefore the one whose entry has the smallest absolute value. Note that the dark pixels shown in Figure 4.4 are identical those in Figure 4.1.

The same idea works generally. In the *arithmetic*[1] *array* $A(v, -w)$, the number

$$vx - wy$$

occurs in row x and column y, where x ranges from 0 to w and y ranges from 0 to v. To represent the segment joining $(0,0)$ and (w,v), the computer should darken the pixel with the smallest absolute value in each column of the arith-

[1]This adjective is pronounced AIR-ITH-MEH-TIK with accents on the first and third syllables.

metic array $A(v, -w)$. The number 0 occurs both in position $(0, 0)$ in the lower left corner and in position (w, v) in the upper right corner of the array, which guarantees that the two endpoints of the segment are dark.

The numbers in the arithmetic array $A(v, -w)$ increase by v as we read across each row, forming an *arithmetic progression* with difference v. As we read up each column, we find an arithmetic progression with difference $-w$. The arithmetic progressions make the entries in the arithmetic array easy to compute by hand and create other patterns we explore later.

A Staircase Shortcut

It is not necessary to compute every entry in the arithmetic array to find the desired pixels. Within each column, the smallest entry in absolute value is the unique[2] number $n = vx - wy$ satisfying

$$-\frac{w}{2} \leq n < \frac{w}{2}. \tag{4.2}$$

We generate the segment from $(0, 0)$ to (w, v) one pixel at a time, working column by column from left to right. Suppose a particular pixel (x, y) has just been darkened, and let the corresponding entry of the arithmetic array be $n = vx - wy$. Because the slope of our segment is at most 1, either pixel $(x + 1, y)$ or pixel $(x + 1, y + 1)$ should be darkened in the next column. The corresponding entries in the arithmetic array are $n + v$ and $n + v - w$, and we only need to decide which of these two numbers satisfies (4.2). This means we can restrict our attention to the entries on a staircase configuration in the arithmetic array, as in Figure 4.5.

Table 4.1 lists successive pixels and values of n for the segment joining $(0, 0)$ and $(13, 8)$. The output agrees with

[2]Two pixels in a column might have values $-w/2$ and $w/2$; we have adopted the arbitrary convention of choosing the pixel with value $-w/2$.

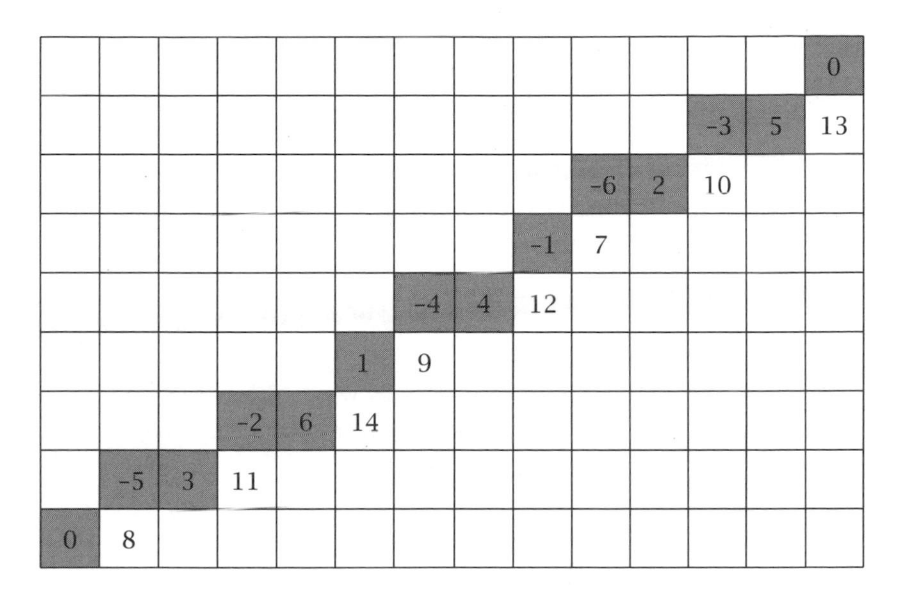

Figure 4.5: A staircase of pixels in the arithmetic array $A(8, -13)$

Figure 4.4. Algorithm 4.2 implements the staircase method in general. The key test for darkening a pixel occurs in step 5, and the value of n is updated in steps 4 and 6.

Table 4.1: Pixels darkened by the line drawing algorithm

(x, y)	$(0,0)$	$(1,1)$	$(2,1)$	$(3,2)$	$\cdots$	$(13,8)$
n	0	-5	3	-2	$\cdots$	0
$\tilde{n} = 2n - 13$	-13	-23	-7	-17	$\cdots$	-13

The shaded cells in the arithmetic array $A(8, -13)$ in Figure 4.5 contain the numbers

$$0,\ -5,\ 3,\ -2,\ 6,\ 1,\ -4,\ 4,\ -1,\ -6,\ 2,\ -3,\ 5,\ 0.$$

Algorithm 4.2. A line drawing algorithm

Input: integers v and w with $0 < v \leq w$

Output: $w + 1$ dark pixels that represent the line segment joining $(0, 0)$ to (w, v)

1. Start at pixel $(x, y) = (0, 0)$. Let $n = 0$.
2. Darken pixel (x, y).
3. If $x = w$, then all $w + 1$ pixels are dark. Stop.
4. Replace (x, y) by $(x + 1, y)$ and n by $n + v$.
5. If $-w/2 \leq n < w/2$, return to step 2.
6. Replace (x, y) by $(x, y + 1)$ and n by $n - w$.
   Return to step 2.

Each number from -6 to 6 occurs once, except for 0, which occurs twice. Also, the *absolute sequence*, obtained by ignoring the negative signs, is a *palindrome*, reading the same backward and forward. The same properties hold for the shaded cells in the arithmetic array $A(v, -w)$ whenever v and w are co-prime. Two positive integers are *co-prime* provided their greatest common divisors is 1. We record these observations for future reference.

Observation. Consider the numbers in the shaded cells of the arithmetic array $A(v, -w)$ for the segment joining $(0, 0)$ and (w, v), where v and w are co-prime and $v < w$.

(a) Each integer n satisfying $-w/2 \leq n < w/2$ occurs once, except for 0, which occurs twice.

(b) The absolute sequence of numbers is palindromic.

4.4 Bresenham's Algorithm

Algorithm 4.2 can be improved. It is not difficult to see that the inequality $-w/2 \leq n$ in step 5 is never violated. Also, the condition $n < w/2$ is equivalent to $2n - w < 0$. If we introduce the parameter $\tilde{n} = 2n - w$, then the algorithm only needs to check the condition $\tilde{n} < 0$ in step 5. (See the last row of Table 4.1.) A computer can check this simple sign condition quickly. The resulting improvements produce Bresenham's famous line drawing algorithm (Algorithm 4.3). The algorithm selects the best pixels to represent a segment while relying almost entirely on addition and subtraction of integers.

Algorithm 4.3. Bresenham's line drawing algorithm

Input: integers v and w with $0 < v \leq w$
Output: $w + 1$ dark pixels that represent the line segment joining $(0, 0)$ and (w, v)

1. Start at pixel $(x, y) = (0, 0)$. Let $\tilde{n} = -w$.
2. Darken pixel (x, y).
3. If $x = w$, then all $w + 1$ pixels are dark. Stop.
4. Replace (x, y) by $(x + 1, y)$ and $\tilde{n}$ by $\tilde{n} + 2v$.
5. If $\tilde{n} < 0$, return to step 2.
6. Replace (x, y) by $(x, y + 1)$ and $\tilde{n}$ by $\tilde{n} - 2w$. Return to step 2.

Our statement of Bresenham's algorithm emphasizes its logical structure. Practical computer implementations in modern programming languages execute steps 2–5 in a loop with x running from 0 to w. Also, the multiplication by 2 in steps 4 and 6 can be eliminated by defining the parameters $2v$ and $2w$ before entering the loop.

4.5 A Touch of Gray: Antialiasing

The smoothness of a line segment on a computer monitor can often be enhanced by *antialiasing,* a process that assigns varying shades of gray to pixels near the segment. The number of shades available is usually a power of 2. Figure 4.6 shows eight shades of gray, ranging from 0 (black)

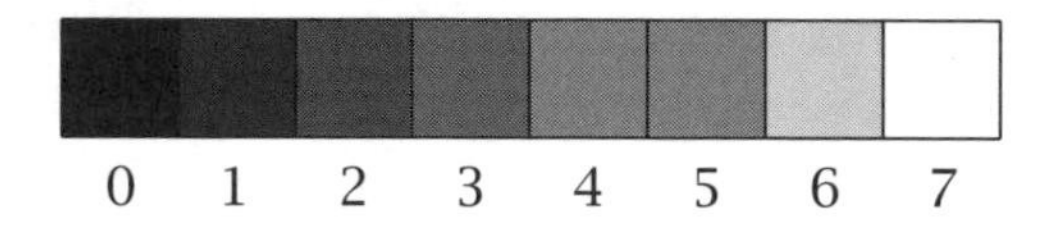

Figure 4.6: Eight shades of gray

to 7 (white), and Figure 4.7 shows the smoother appearance of an antialiased line segment using these eight shades.

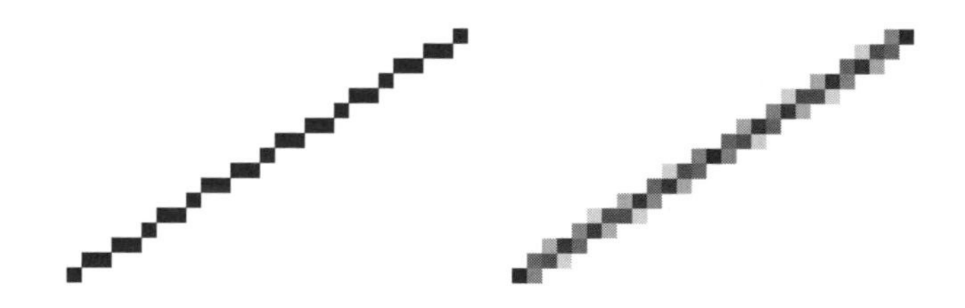

Figure 4.7: The antialiased segment is on the right

We outline one scheme to carry out antialiasing. Assign shade 0 (black) to a pixel if the line segment passes directly through its center. If the segment passes between two pixel centers in a column, then assign the shades of gray in rough proportion to the distances of these two pixels from the segment. In particular, we can assign shade s to one pixel and shade $7 - s$ to the other one for some value of $s = 1, 2, \ldots, 6$. All other pixels remain white ($s = 7$). Because the distance between a line and the pixel (x, y) is determined by the parameter $n = vx - wy$, Bresenham's algorithm can be readily modified to produce antialiased lines.

4.6 Leap Years and Line Drawing

A perplexing mathematical question arises in the design of calendars. How do we impose the familiar discrete divisions (day, month, year) on the continuous, periodic movements of the earth, moon, and sun? Any answer to this question entails compromise since the exact periods of the celestial bodies cannot be expressed as the ratios of small integers. Truly accurate calendars are necessarily complex, while simple calendars are inaccurate. Calendars rely on approximations that sacrifice some degree of accuracy for the sake of simplicity and predictability. Many calendars insert a *leap day* or *month* in specified *leap years*—a process called *intercalation*—to correct for the accumulation of approximation errors and realign the dates with natural phenomena. Some calendars distribute the leap days and months evenly throughout some cycle of years.

We now discuss a connection between the computer line drawing problem and the leap year patterns in two calendars.

The Hebrew Calendar

The Hebrew calendar is an ancient lunar calendar with an intercalary month. It is still used today to determine the dates of religious observances. Every 19-year cycle contains 11 common years (with 12 months) and 7 leap years (with 13 months). Within the cycle, the leap years occur in years

3, 6, 8, 11, 14, 17, and 19.

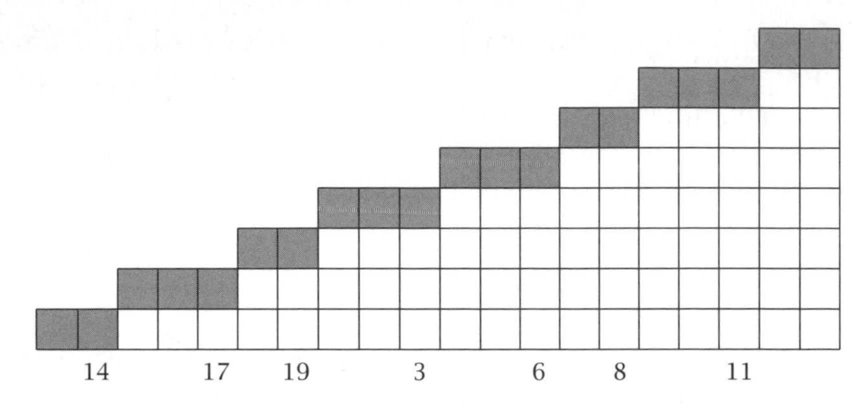

Figure 4.8: The 19-year cycle of leap years in the Hebrew calendar

The leap year pattern can be found in the configuration of pixels representing the segment from $(0,0)$ to $(19,7)$. The dark pixels in Figure 4.8 distribute a rise of 7 vertical units (the leap years) as evenly as possible across a horizontal run of 19 years. Year x is a leap year when pixel (x, y) is followed by a step up to pixel $(x+1, y+1)$. If we let pixels $(0,0)$ and $(19,7)$ correspond to year 13 of the 19-year cycle, then the cyclic pattern of 2- and 3-unit horizontal runs of pixels agrees with the pattern of leap years in the Hebrew calendar.

The Arithmetical Islamic Calendar

The lunar calendar of the Islamic world depends on local sightings of the crescent moon and therefore on the prevailing weather conditions, the visual acumen of an observer, and small changes in the orbits of the earth and moon. Islamic scholars of the eighth century devised the *arithmetical Islamic calendar*, which uses simple mathematical rules to assign dates that are remarkably close to the correct ones. A common year has 354 days, and a leap year has 355 days in the arithmetical calendar. The 11 leap years occur in years

$$2, 5, 7, 10, 13, 16, 18, 21, 24, 26, \text{and } 29$$

of a 30-year cycle. The leap year pattern corresponds to the horizontal runs of pixels representing the segment joining $(0, 0)$ and $(30, 11)$. See Figure 4.9. Pixels $(0, 0)$ and $(30, 11)$ correspond to year 12 of the 30-year cycle.

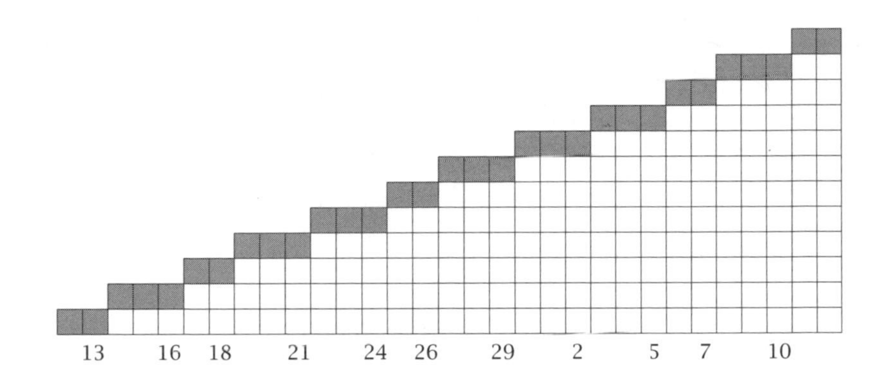

Figure 4.9: Leap years in the arithmetical Islamic calendar

Despite its mathematical virtues, the arithmetical calendar is not widely used. It has deviated at times from the observation-based calendar by one or two days, an unacceptable shortcoming for religious purposes.

A Direct Line of Reasoning

How did people concoct leap year patterns centuries before computer line drawing problems had been contemplated? Here is one plausible path. In the Hebrew calendar, each common year is 7/19 of a month too short. It is reasonable to add one full leap month when the accumulated error exceeds 1/2 month. After the first year, the accumulated error is 7/19. After the second year, the error is 14/19, which is greater than 1/2. So a leap month is inserted in the second year, and the accumulated error is adjusted to $(14 - 19)/19 = -5/19$. After the third and fourth years, the accumulated errors are 2/19 and 9/19, respectively. After the fifth year, the error is 16/19, so a leap month is inserted and the error is adjusted to $-3/19$. Continuing the process yields the full pattern of leap years in the Hebrew calendar.

The method of accumulated errors can also be applied directly to the line drawing problem. Not surprisingly, it leads to Bresenham's algorithm. In fact, this is the usual method of deriving Bresenham's algorithm. The geometric distances measured by the entries in our arithmetic arrays are visual manifestations of the accumulated errors.

4.7 Diophantine Approximations

We turn from from the geometric problem of approximating lines by pixels to the numerical one of approximating real numbers by fractions. We will see that the two problems are closely related.

Two of the most famous approximations in mathematics are

$$\pi \approx \frac{22}{7} \qquad \text{and} \qquad \pi \approx 3.14.$$

The first approximation is superior to the second in both simplicity and accuracy. The fraction 22/7 is simpler than $3.14 = 157/50$ because the denominator 7 is less than the denominator 50. To compare the accuracies, we look at the absolute errors

$$\begin{aligned} \left|\pi - \frac{22}{7}\right| &= 0.001264\ldots \\ \text{and} \qquad |\pi - 3.14| &= 0.001592\ldots. \end{aligned}$$

A *Diophantine approximation* of the real number u is a fraction y/x chosen to meet the two conflicting goals of simplicity and accuracy. We are given u and want to find a fraction y/x with a small denominator x and an absolute error

$$\left| u - \frac{y}{x} \right|$$

close to 0. Because the numerator y and the denominator x must be integers, we are approximating u by a *rational number*. Diophantine approximations are named for Diophantus of Alexandria, a third-century mathematician whose study of rational solutions to equations played a major role in the development of algebra and number theory.

Suppose the denominator x in a Diophantine approximation is specified in advance. The multiples of $1/x$ are equally spaced points on a number line, and Figure 4.10 makes it clear than any given real number u must fall within $1/2x$ of the nearest multiple of $1/x$. It follows that for each denominator x, there is a numerator y such that

$$\left| u - \frac{y}{x} \right| \le \frac{1}{2x}.$$

Figure 4.10: The number u is within $1/2x$ of a multiple of $1/x$

Better Diophantine approximations are available if the denominator x is not specified in advance. The elementary error bound $1/2x$ can often be replaced by the much smaller expression $1/2x^2$. An important result of this type deals with approximations to *irrational numbers*, numbers like π and $\sqrt{2}$ that cannot be expressed as the ratio of two integers.

Diophantine approximations to irrational numbers. An irrational number u has infinitely many Diophantine approximations

$$u \approx \frac{y}{x} \quad \text{with absolute error} \quad \left| u - \frac{y}{x} \right| < \frac{1}{2x^2}.$$

Approximations of the type described in the preceding result include

$$\pi \approx \frac{22}{7} \text{ with absolute error } < \frac{1}{2(7)^2} = 0.01\ldots$$

and

$$\pi \approx \frac{355}{113} \text{ with absolute error } < \frac{1}{2(113)^2} = 0.00039\ldots.$$

Diophantine approximations to irrational numbers are best found by examining *continued fractions*, a fascinating topic we do not explore here.

Diophantine Approximations and Line Drawing

Now let us look at Diophantine approximations y/x to a rational number

$$u = \frac{v}{w}.$$

Of course, the error is 0 if we choose $y/x = v/w$, but we insist on using a denominator x that is strictly less than w. In other words, we seek to approximate one fraction by another one with a smaller denominator. Approximations of this type are implicit in the choice of leap year frequencies in calendars. For instance, astronomical measurements have shown that the average number of days between new moons is about 29.530589, an unwieldy rational number. The accuracy of the arithmetical Islamic calendar is a consequence of the excellent Diophantine approximation

$$29.530589 \approx \frac{10631}{360}.$$

The source of the mysterious fraction 10631/360 is given in Problem 11. The tiny absolute error

$$\left|29.530589 - \frac{10631}{360}\right| = 0.000033\ldots$$

reveals the sophistication of early Islamic astronomy. The Julian calendar[3] used widely in Europe until the late sixteenth century had an error roughly 20 times as large.

Diophantine approximations to rational numbers are related to the computer line drawing problem. If we approximate the fraction $u = v/w$ by the fraction y/x, the absolute error is

$$\left|\frac{v}{w} - \frac{y}{x}\right| = \frac{|vx - wy|}{wx} = \frac{d_{\text{vert}}}{x},$$

where d_{vert} is the vertical distance between the point (x, y) and the line joining the points $(0, 0)$ and (w, v). To make the error small, we should choose the point (x, y) close to this line—the same task we faced in the computer line drawing problem. When $v < w$, the arithmetic array $A(v, -w)$ contains the information we need.

Example 1. To approximate the fraction $8/13$, we inspect the arithmetic array $A(8, -13)$ in Figure 4.4. The numbers 1 and -1 appear in positions $(x, y) = (5, 3)$ and $(8, 5)$, giving us the Diophantine approximations

$$\frac{8}{13} \approx \frac{3}{5} \quad \text{and} \quad \frac{8}{13} \approx \frac{5}{8}$$

with absolute errors

$$\left|\frac{8}{13} - \frac{3}{5}\right| = \frac{1}{13 \times 5} = 0.015\ldots$$

$$\text{and} \quad \left|\frac{8}{13} - \frac{5}{8}\right| = \frac{1}{13 \times 8} = 0.009\ldots.$$

[3] A solar calendar with one leap year every four years.

In general, to approximate the fraction v/w, we locate the positions (x, y) of 1 and -1 in the array $A(v, -w)$. Our observations in Section 4.3 guarantee that one of these two numbers occurs in the left half of the array and thus satisfies $x \le w/2$. This gives us a Diophantine approximation

$$\frac{v}{w} \approx \frac{y}{x}$$

with absolute error

$$\left|\frac{v}{w} - \frac{y}{x}\right| = \frac{1}{wx} \le \frac{1}{(2x)x} = \frac{1}{2x^2}.$$

Here is a formal statement.

Diophantine approximations to rational numbers. Let $u = v/w$ be a rational number in lowest terms with $0 < v < w$. Then u has a Diophantine approximation

$$u \approx \frac{y}{x} \quad \text{with absolute error} \quad \left|u - \frac{y}{x}\right| \le \frac{1}{2x^2}$$

and denominator x at most $w/2$.

Euclid's Algorithm

There is no need to inspect arithmetic arrays to produce the rational Diophantine approximation given above. A better approach is based on the familiar algorithm of Euclid for finding the greatest common divisor of two positive numbers.

To find the greatest common divisor of v and w, Euclid's algorithm divides v into w to produce a quotient q and remainder r. The relation

$$w = vq + r$$

makes it clear that any common divisor of v and w is also a divisor of r. In fact,

$$\gcd(w, v) = \gcd(v, r),$$

and the procedure can be repeated with r and v in place of v and w. Continue in the same manner until a remainder of 0 occurs. The last nonzero remainder is the greatest common divisor of the original numbers v and w.

Example 2. When $v = 103$ and $w = 127$, Euclid's algorithm gives

$$\begin{aligned} \gcd(127, 103) &= \gcd(103, 24) = \gcd(24, 7) \\ &= \gcd(7, 3) = \gcd(3, 1) = \gcd(1, 0) \\ &= 1. \end{aligned}$$

See Table 4.2 for details. We find that $\gcd(127, 103) = 1$, and so the numbers 127 and 103 are co-prime.

Table 4.2: Euclid's algorithm and the gcd of 127 and 103

$127 \div 103$	=	1	with remainder 24
$103 \div 24$	=	4	with remainder 7
$24 \div 7$	=	3	with remainder 3
$7 \div 3$	=	2	with remainder 1
$3 \div 1$	=	3	with remainder 0

For the purposes of our Diophantine approximations, it is better to express the successive remainders in the algorithm in terms of $v = 103$ and $w = 127$, as in Table 4.3. The last row of the table tells us that the equation

$$103x + 127y = 1$$

is satisfied by $x = 37$ and $y = -30$.

In general, Euclid's algorithm leads to the following result.

Table 4.3: Euclid's algorithm and the co-primality theorem

$$
\begin{aligned}
[127, 103] &= [w, v] \\
[103, 24] &= [v, w - v] \\
[24, 7] &= [w - v, v - 4(w - v)] \\
&= [w - v, -4w + 5v] \\
[7, 3] &= [-4w + 5v, (w - v) - 3(-4w + 5v)] \\
&= [-4w + 5v, 13w - 16v] \\
[3, 1] &= [13w - 16v, (-4w + 5v) - 2(13w - 16v)] \\
&= [13w - 16v, -30w + 37v]
\end{aligned}
$$

Co-primality theorem. If v and w are co-prime, then there are integers x and y that satisfy $vx + wy = 1$.

The co-primality theorem lets us show that $|y|/|x|$ is a good Diophantine approximation to the fraction v/w.

Example 3. To find a good Diophantine approximation to the fraction 103/127, we begin with the relation

$$103 \times 37 \ + \ 127 \times (-30) \ = \ 1$$

from the co-primality theorem. Divide both sides by the product 127×37 and take the absolute value to see that

$$\left| \frac{103}{127} - \frac{30}{37} \right| = \frac{1}{127 \times 37} \leq \frac{1}{2(37)^2}.$$

This inequality gives the Diophantine approximation

$$\frac{103}{127} \approx \frac{30}{37} \quad \text{with absolute error } \leq \frac{1}{2(37)^2}.$$

4.8 Notes and References

Bresenham describes his algorithm in [2], which is also available online at the link given. The book [3] discusses Bresenham's algorithm and connections to other problems in computer graphics; computer code is included. Wu and Rokne [11] describe a variant of Bresenham's algorithm that darkens pixels two at a time.

The mathematics of calendars is covered in detail in [7] and [8]. The link between leap years and computer line drawing is discussed by Harris and Reingold [4]. They also relate computer line drawing to Euclid's algorithm.

See [10] for a discussion of Diophantus' role in the history of mathematics. Diophantine approximations are treated in [5] and in many books on number theory, including the one by Ogilvy and Anderson [6]. For more on continued fractions, see the last three chapters of [1]. A good introduction to Euclid's algorithm is found in Stillwell's book [9].

1. Ball, Keith, *Strange Curves, Counting Rabbits, and Other Mathematical Explorations.* Princeton University Press, Princeton, New Jersey, 2003.
2. Bresenham, J. E., Algorithm for computer control of a digital plotter, *IBM Systems Journal* **4** no. 1 (1965), 25–30. Also available at: `www.research.ibm.com/journal/sj/041/ibmsjIVRIC.pdf`
3. Glassner, Andrew S. (ed.), *Graphics Gems.* Academic Press Professional, Boston, 1990.
4. Harris, M. A., and Reingold, E. M., Line drawing, leap years, and Euclid, *ACM Computing Surveys* **36** (2004), 68–80.
5. Niven, Ivan, *Diophantine Approximations.* Dover, New York, 2008.
6. Ogilvy, C. Stanley, and Anderson, John T., *Excursions in Number Theory.* Dover, New York, 1988.
7. Reingold, Edward M., and Dershowitz, Nachum, *Calendrical Calculations: The Millennium Edition.* Cambridge University Press, Cambridge, U.K., 2001.
8. Reingold, Edward M., and Dershowitz, Nachum, *Calendrical Tabulations, 1900–2200.* Cambridge University Press, Cambridge, U.K., 2002.
9. Stillwell, John, *Elements of Number Theory.* Springer, New York, 2003.
10. Stillwell, John, *Numbers and Geometry.* Springer, New York, 1998.
11. Wu, X., and Rokne, J. G., Double-step incremental generation of lines and circles, *Computer Vision, Graphics, and Image Processing* **37** (1987), 331–334.

4.9 Problems

1. (a) Use a suitable pair of similar right triangles in Figure 4.3 to show that

$$\frac{v}{w} = \frac{d_{\text{vert}} + y}{x}.$$

(b) Use part (a) to deduce the vertical distance formula.

(c) Use another pair of similar right triangles in Figure 4.3 to deduce that

$$d = \frac{|vx - wy|}{\sqrt{v^2 + w^2}}.$$

2. Use the arithmetic array $A(3, -5)$ in Figure 4.11 to darken the six pixels for the segment joining $(0, 0)$ and $(5, 3)$.

-15	-12	-9	-3	-4	0
-10	-7	-4	-1	2	5
-5	-2	1	4	7	10
0	3	6	9	12	15

Figure 4.11: The arithmetic array $A(3, -5)$

3. Use the arithmetic array $A(5, -8)$ in Figure 4.12 to darken the nine pixels for the segment joining $(0, 0)$ and $(8, 5)$.

-40	-35	-30	-25	-20	-15	-10	-5	0
-32	-27	-22	-17	-12	-7	-2	3	8
-24	-19	-14	-9	-4	1	6	11	16
-16	-11	-6	-1	4	9	14	19	24
-8	-3	2	7	12	17	22	27	32
0	5	10	15	20	25	30	35	40

Figure 4.12: The arithmetic array $A(5, -8)$

4. If Algorithm 4.2 is used with $v = w = 5$ to represent the segment from $(0,0)$ to $(5,5)$, what is the value of n each time step 2 is executed?

5. What are the omitted entries in Table 4.1?

6. Let Bresenham's algorithm be used to represent the segment from $(0,0)$ to (w,v), where w is odd. Show that pixel (x,y) is dark if and only if pixel $(w-x, v-y)$ is dark.

7. Use the result of the Problem 6 to modify Bresenham's algorithm so that it darkens pixels from both ends of the segment simultaneously. Either assume w is odd, or else make a small adjustment to the conclusion of Problem 6 in case w is even.

8. Verify that the successive numerators 7, -5, 2, 9, -3, ... of the accumulated errors for the Hebrew calendar occur in the positions of the dark pixels for the segment joining $(0,0)$ and $(19,7)$ in the arithmetic array $A(7,-19)$.

9. (a) Which pixels are darkened by Bresenham's algorithm for the segment joining $(0,0)$ and $(12,5)$?

   (b) Use the answer to part (a) to distribute five months of 31 days and seven months of 30 days over a 365-day year in some reasonable manner.

10. The arithmetical Islamic calendar described by Figure 4.9 has a variant in which a leap year in year 15 replaces the one in year 16. What reasonable modification of our convention in (4.2) agrees with this variant?

11. In the arithmetical Islamic calendar, the 12 months in a year alternate between 30 and 29 days, except that the twelfth month has 30 days in a leap year. Show that there are $10631/360$ days per month on average.

12. (a) What calendar is related to the Diophantine approximation

$$365.2422 \approx 365 + \frac{1}{4}?$$

(b) What calendar is related to the Diophantine approximation

$$365.2422 \approx 365 + \frac{97}{400}?$$

13. Sketch an antialiased representation of the segment joining $(0,0)$ and $(13,8)$ using four shades of gray. Compare your sketch to Figure 4.1.

14. Find a Diophantine approximation

$$\frac{355}{113} \approx \frac{y}{x}$$

with $x < 113$ and absolute error at most $1/2x^2$.

15. Find a Diophantine approximation

$$\frac{55}{89} \approx \frac{y}{x}$$

with $x < 89$ and absolute error at most $1/2x^2$.

God writes straight with crooked lines.
SPANISH PROVERB

Truth is much too complicated
to allow anything but approximations.
JOHN VON NEUMANN

5

Measure Water with a Vengeance

On the fountain there should be two jugs.
Do you see them?
SIMON GRUBER

5.1 Simon Says: Measure Water

In the 1995 action movie *Die Hard: With a Vengeance,* Bruce Willis plays a maverick law enforcement officer who must solve a series of fiendish puzzles posed by Simon Gruber, the mastermind of a diabolical bank heist. In one memorable scene, Simon directs Willis and his reluctant sidekick to a fountain in a public park, where they find two empty, unmarked jugs with capacities 3 and 5 gallons and a command to measure precisely 4 gallons of water to disarm a bomb. The 4 gallons must be measured by pouring water from the fountain into and between the two jugs. After a tense discussion, the pair succeeds, although the movie's editing omits several steps of the process.

Perhaps you have seen similar puzzles, but with jugs of different capacities. Here is the basic problem of this chapter.

Water measuring problem. We are at a fountain with two unmarked jugs with capacities v and w units. Can we measure exactly n units of water? If so, how?

The jug capacities are now measured in some unspecified units of volume rather than gallons. Bruce Willis faced the case $v = 3$, $w = 5$, $n = 4$.

Legal Moves

There are three types of legal moves in a water measuring problem:

- Fill a jug with water from the fountain.
- Empty a jug by pouring its contents into the fountain.
- Pour water from one jug to the other until either the first jug is empty or the second one is full.

We assume that the pouring and measuring processes are carried out perfectly. No evaporation occurs, water does not adhere to the inside of a jug when we empty it, and so on.

Although we generally applaud the ability to think "outside the box" (or outside the jug), our rules forbid the creative solution to the Bruce Willis problem illustrated in Figure 5.1, in which both jugs are filled and then tilted until water just touches the rims and bottoms. The portion of each jug occupied by water has the same shape and volume as the unoccupied portion. This means that each jug is half-filled with water, and the total amount of water is $3/2 + 5/2 = 4$ gallons. This clever approach requires a steady hand and is only available if the jugs possess enough symmetry.

Co-prime Capacities

In our water measuring problem, we assume that the jug capacities v and w are integers, as is n, the desired amount of water. It is also natural to assume that $0 \leq n \leq v + w$. Because it is easy to measure 0 units of water (keep both

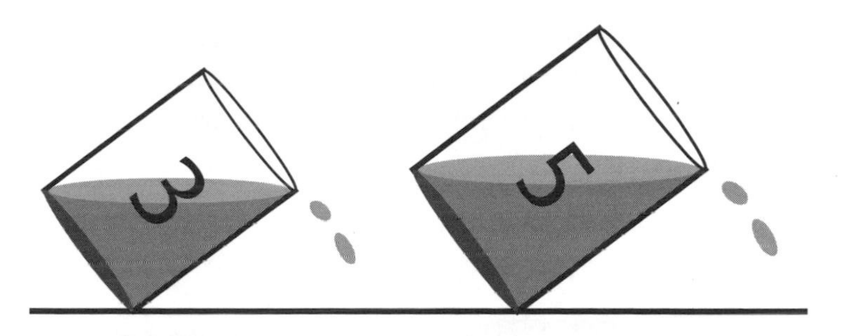

Figure 5.1: Each jug is half-filled with water

jugs empty) and to measure $v + w$ units (fill both jugs), most of our effort is devoted to the cases $n = 1, 2, \ldots, v + w - 1$.

If both jug capacities are divisible by 4, then it is not difficult to see that the only measurable amounts of water are also divisible by 4. We might as well divide through by 4 in all of our measurements. This division is tantamount to changing the units, say, from quarts to gallons, and does not alter the essential nature of the water measuring problem. Two positive integers are *co-prime* provided their greatest common divisor (gcd) is 1. We always assume that our given jug capacities v and w are co-prime, and this is without loss of generality.

Every water measuring problem satisfying our reasonable assumptions can be solved. In fact, we will discuss three solution methods. The first produces a correct sequence of pourings by tracing the zigzag path of a bouncing billiard ball on an unusual billiard table. The second method is algorithmic, relying on a few simple rules that are easy to remember and follow. The third method uses a staircase of numbers in a special rectangular array to generate the pouring instructions.

It will be apparent that our three solutions are in some sense equivalent to one another. They lead to identical sequences of pouring instructions for any given water measuring problem. However, the solutions offer different perspectives. For instance, the rectangular arrays in the third solution are related to one of Fermat's famous results in number theory.

5.2 A Recipe for Bruce Willis

Suppose you have devised a sequence of legal moves to measure a desired amount of water. There are several ways to present your solution.

Words and Pictures

The most natural way to give an answer to a water measuring problem is through a recipe of instructions in complete sentences. For instance, Algorithm 5.1 is a recipe for the Bruce Willis problem. A sequence of pictures would also do the job. The first picture would show both jugs empty, the second would show the smaller jug full, and the larger jug empty, and so on.

States

Some notation allows us to describe our water measuring activities succinctly. A *state* is a description of the amount of water in each jug at a particular time. We let

$$\lfloor a \mid b \rfloor \quad = \quad \text{the state with } a \text{ units in the smaller jug and } b \text{ units in the larger jug.}$$

The notation serves as a small picture of the two jugs side by side. The smaller jug always appears on the left. Now we can describe the solution to a water measuring problem by listing the states in order.

Algorithm 5.1. A recipe for the Bruce Willis problem

Ingredients: fountain water; 3- and 5-gallon jugs
Goal: 4 gallons of water

1. Fill the 3-gal. jug from the fountain.
2. Empty the 3-gal. jug to the 5-gal. jug.
3. Fill the 3-gal. jug from the fountain.
4. Fill the 5-gal. jug from the 3-gal. jug.
   The 3-gallon jug contains 1 gallon.
5. Empty the 5-gal. jug to the fountain.
6. Pour the 1 gal. of water into the 5-gal. jug.
7. Fill the 3-gal. jug from the fountain.
8. Empty the 3-gal. jug to the 5-gal. jug.
   The 5-gallon jug contains 4 gallons.

Example 1. (a) The recipe for the Bruce Willis problem in Algorithm 5.1 takes the compact form

$$\lfloor 0 \,|\, 0 \rfloor \sim\!\sim \lfloor 3 \,|\, 0 \rfloor \sim\!\sim \lfloor 0 \,|\, 3 \rfloor \sim\!\sim \lfloor 3 \,|\, 3 \rfloor \sim\!\sim \lfloor 1 \,|\, 5 \rfloor$$

$$\sim\!\sim \lfloor 1 \,|\, 0 \rfloor \sim\!\sim \lfloor 0 \,|\, 1 \rfloor \sim\!\sim \lfloor 3 \,|\, 1 \rfloor \sim\!\sim \lfloor 0 \,|\, 4 \rfloor .$$

The ripples between states signify the pouring of water. There are eight pourings in this solution.

(b) A different solution with just seven pourings starts by filling the 5-gallon jug:

$$\lfloor 0 \,|\, 0 \rfloor \sim\!\sim \lfloor 0 \,|\, 5 \rfloor \sim\!\sim \lfloor 3 \,|\, 2 \rfloor \sim\!\sim \lfloor 0 \,|\, 2 \rfloor \sim\!\sim \lfloor 2 \,|\, 0 \rfloor$$

$$\sim\!\sim \lfloor 2 \,|\, 5 \rfloor \sim\!\sim \lfloor 3 \,|\, 4 \rfloor \sim\!\sim \lfloor 0 \,|\, 4 \rfloor .$$

5.3 Skew Billiard Tables

The skew billiard table in Figure 5.2 solves the Bruce Willis problem geometrically. The table is a 3-by-5 parallelogram with 60° and 120° angles and no pockets. The grid-lines partition the table into small equilateral triangles, and the labels for 16 different states occur on the boundary. At least one jug is empty or full in each of these states.

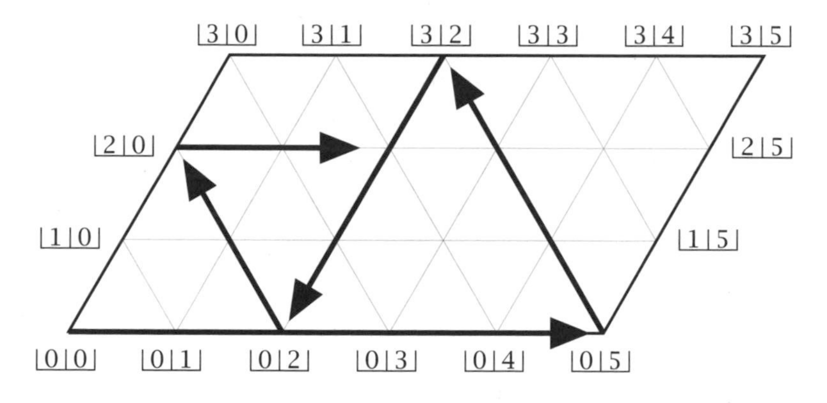

Figure 5.2: The bouncing ball tells us how to measure water

Place a billiard ball at the corner $\lfloor 0|0 \rfloor$. Then set the ball in motion along the lower edge and watch it bounce around the skew table, as indicated by the arrows. The ball always travels along grid-lines with each bounce creating a 60° angle. Successive labels give a sequence of states to solve the water measuring problem. The ball starts at $\lfloor 0|0 \rfloor$ and first hits the corner $\lfloor 0|5 \rfloor$. Then it bounces up to $\lfloor 3|2 \rfloor$, down to $\lfloor 0|2 \rfloor$, up to $\lfloor 2|0 \rfloor$, over to $\lfloor 2|5 \rfloor$, and so on. Eventually, the ball arrives at the desired target state $\lfloor 0|4 \rfloor$, and we have measured 4 units of water, as in Example 1(b).

If we allow the ball to continue bouncing beyond $\lfloor 0|4 \rfloor$, then we obtain a *complete state sequence*

$$\lfloor 0 \mid 0 \rfloor \rightsquigarrow \lfloor 0 \mid 5 \rfloor \rightsquigarrow \lfloor 3 \mid 2 \rfloor \rightsquigarrow \lfloor 0 \mid 2 \rfloor \rightsquigarrow \lfloor 2 \mid 0 \rfloor$$
$$\rightsquigarrow \lfloor 2 \mid 5 \rfloor \rightsquigarrow \lfloor 3 \mid 4 \rfloor \rightsquigarrow \lfloor 0 \mid 4 \rfloor \rightsquigarrow \lfloor 3 \mid 1 \rfloor \rightsquigarrow \lfloor 0 \mid 1 \rfloor$$
$$\rightsquigarrow \lfloor 1 \mid 0 \rfloor \rightsquigarrow \lfloor 1 \mid 5 \rfloor \rightsquigarrow \lfloor 3 \mid 3 \rfloor \rightsquigarrow \lfloor 0 \mid 3 \rfloor \rightsquigarrow \lfloor 3 \mid 0 \rfloor.$$

When the ball finally arrives at the corner $\lfloor 3 \mid 0 \rfloor$, it has visited each state once, except for the uninteresting corner state $\lfloor 3 \mid 5 \rfloor$. The ball has produced pouring instructions to measure n units of water with the two given jugs for $n = 0, 1, \ldots, 7$.

Abbreviated Solutions

The complete state sequence just given may be abbreviated as

$$0 \rightsquigarrow 5 \rightsquigarrow 2 \rightsquigarrow 7 \rightsquigarrow 4 \rightsquigarrow 1 \rightsquigarrow 6 \rightsquigarrow 3.$$

We replaced the state $\lfloor a \mid b \rfloor$ by the single number $a + b$ (the total amount of water in the two jugs at each step) and deleted duplications. You can verify that the resulting *abbreviated state sequence* provides enough information to determine all necessary pourings and reconstruct a complete state sequence.

If we set the billiard ball in motion along the left edge of the table initially (Figure 5.3), then we generate the complete state sequence

$$\lfloor 0 \mid 0 \rfloor \rightsquigarrow \lfloor 3 \mid 0 \rfloor \rightsquigarrow \lfloor 0 \mid 3 \rfloor \rightsquigarrow \lfloor 3 \mid 3 \rfloor \rightsquigarrow \lfloor 1 \mid 5 \rfloor$$
$$\rightsquigarrow \lfloor 1 \mid 0 \rfloor \rightsquigarrow \lfloor 0 \mid 1 \rfloor \rightsquigarrow \lfloor 3 \mid 1 \rfloor \rightsquigarrow \lfloor 0 \mid 4 \rfloor \rightsquigarrow \lfloor 3 \mid 4 \rfloor$$
$$\rightsquigarrow \lfloor 2 \mid 5 \rfloor \rightsquigarrow \lfloor 2 \mid 0 \rfloor \rightsquigarrow \lfloor 0 \mid 2 \rfloor \rightsquigarrow \lfloor 3 \mid 2 \rfloor \rightsquigarrow \lfloor 0 \mid 5 \rfloor$$

with the corresponding abbreviated state sequence

$$0 \rightsquigarrow 3 \rightsquigarrow 6 \rightsquigarrow 1 \rightsquigarrow 4 \rightsquigarrow 7 \rightsquigarrow 2 \rightsquigarrow 5.$$

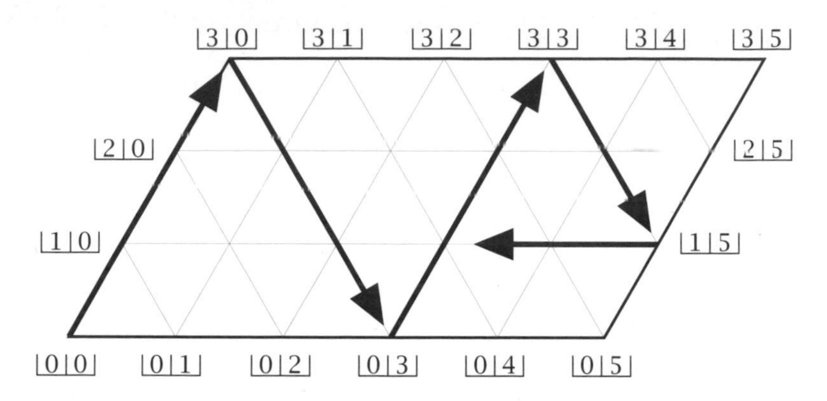

Figure 5.3: The billiard ball can bounce in the other direction, too

The new sequences are essentially the reversals of the earlier ones. Each amount from 0 to 7 units is accounted for exactly once in the abbreviated sequences.

The 15 states and 8 abbreviated states can be arranged in circles, as in Figure 5.4. The figure makes it clear that we can measure any amount from 0 to 7 units with at most seven pourings by correctly choosing either a clockwise or a counterclockwise traversal of the circle.

In summary, the motion of a bouncing billiard ball on a skew 3-by-5 billiard table generates instructions to measure n gallons for $n = 0, 1, \ldots, 7$ using jugs with capacities 3 and 5 gallons.

At the Movies

The water measuring scene in *Die Hard: With a Vengeance* confused some audience members, including the sharp-eyed reviewer Roger Ebert, and with good reason. The solution carried out by Bruce Willis and his sidekick contains an inconsistency—a continuity error in the lingo of movie buffs. The characters arrive at the state ⌊0⌊3⌋, presumably by fill-

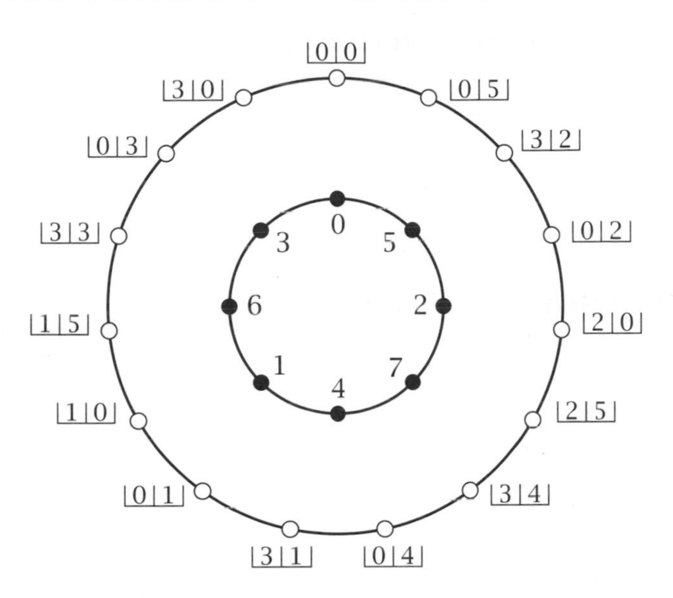

Figure 5.4: Fifteen states and eight abbreviated states

ing the 3-gallon jug and emptying it into the 5-gallon jug. Figure 5.4 tells us that just six more pourings are needed to measure the desired 4 gallons. However, the movie suddenly cuts to the state ⌊2 | 5⌋, and we see that the duo must have backtracked and proceeded the other way around the circle.[1] Cinematic accuracy compels us to note that once the characters reach the state ⌊3 | 4⌋, they ignore the full 3-gallon jug and simply place the jug with 4 gallons on a scale, thereby disarming the bomb.

5.4 Big Problems

The clever skew billiard table idea was introduced in 1939 by M. C. K. Tweedie, who showed that his method works for any two co-prime jug capacities.

[1] Some television broadcasts of the movie edit the scene differently.

Water measuring theorem. We are at a fountain with two empty, unmarked jugs of co-prime capacities v and w units and want to measure n units of water for some $n = 1, 2, \ldots, v + w - 1$.

(a) The motion of a billiard ball on a skew v-by-w table gives a correct sequence of pouring instructions.
(b) At most $v + w - 1$ pourings are required.

To see why (b) is true, note that there are $2v + 2w$ labeled states on the boundary of the v-by-w skew billiard table, and if we exclude the "both jugs full" corner state, then the ball can reach any state in at most $2v + 2w - 1$ bounces, whether we start the ball in motion along the lower edge or along the left edge. Because the two complete state sequences are reversals of one another, the shorter path to any given state has at most $v + w - 1$ bounces.

Example 2. How can we measure 15 units of water using jugs with capacities 7 and 10 units?

We leave you to construct a suitable skew billiard table and produce the following sequence of 19 pourings.

$$\lfloor 0 \mid 0 \rfloor \rightsquigarrow \lfloor 0 \mid 10 \rfloor \rightsquigarrow \lfloor 7 \mid 3 \rfloor \rightsquigarrow \lfloor 0 \mid 3 \rfloor \rightsquigarrow \lfloor 3 \mid 0 \rfloor$$

$$\rightsquigarrow \lfloor 3 \mid 10 \rfloor \rightsquigarrow \lfloor 7 \mid 6 \rfloor \rightsquigarrow \lfloor 0 \mid 6 \rfloor \rightsquigarrow \lfloor 6 \mid 0 \rfloor \rightsquigarrow \lfloor 6 \mid 10 \rfloor$$

$$\rightsquigarrow \lfloor 7 \mid 9 \rfloor \rightsquigarrow \lfloor 0 \mid 9 \rfloor \rightsquigarrow \lfloor 7 \mid 2 \rfloor \rightsquigarrow \lfloor 0 \mid 2 \rfloor \rightsquigarrow \lfloor 2 \mid 0 \rfloor$$

$$\rightsquigarrow \lfloor 2 \mid 10 \rfloor \rightsquigarrow \lfloor 7 \mid 5 \rfloor \rightsquigarrow \lfloor 0 \mid 5 \rfloor \rightsquigarrow \lfloor 5 \mid 0 \rfloor \rightsquigarrow \lfloor 5 \mid 10 \rfloor.$$

The corresponding abbreviated state sequence is

$$0 \rightsquigarrow 10 \rightsquigarrow 3 \rightsquigarrow 13 \rightsquigarrow 6 \rightsquigarrow 16 \rightsquigarrow 9 \rightsquigarrow 2 \rightsquigarrow 12 \rightsquigarrow 5 \rightsquigarrow 15.$$

5.5 How to Measure Water: An Algorithm

Careful examination of the skew billiard table method leads to a general process to solve water measuring problems.

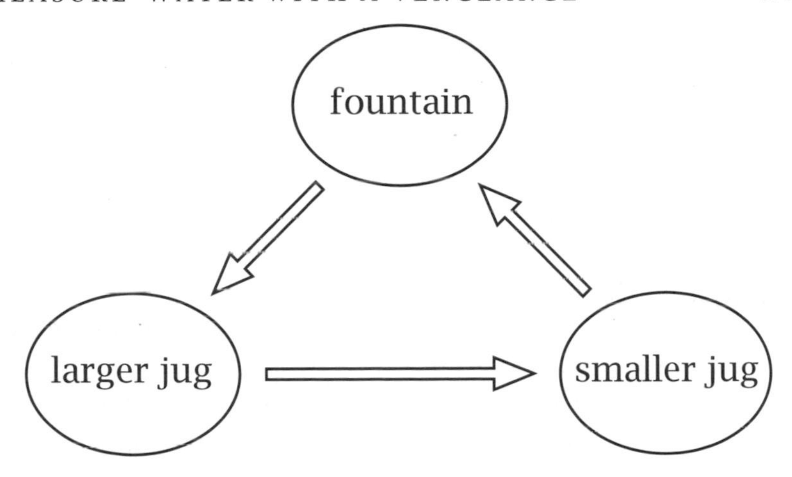

Figure 5.5: A mnemonic for measuring water

The mnemonic in Figure 5.5 makes the process easy to remember. Observe that the billiard ball moves in just three directions in Figure 5.2, corresponding to three types of pouring instructions:

- Fill the larger jug with fountain water.
- Pour water from the larger jug to the smaller.
- Empty the smaller jug to the fountain.

This means we can always solve a water measuring problem by moving water in the cyclic manner indicated in Figure 5.5. In other words, if we arrange the states in a circle, as in Figure 5.4, we can arrive at any state by consistently proceeding in one direction around the circle—unlike Bruce Willis in the movie.

Our discussion is formalized in Algorithm 5.2. Switch the words "larger" and "smaller" throughout to obtain a companion algorithm in which the water moves in the opposite direction.

Algorithm 5.2. Water measuring algorithm for two jugs

Input: two jugs of co-prime capacities v and w units;
fountain water;
positive integer n with $1 \le n \le v + w - 1$.

Output: n units of water in one jug

1. If the larger jug is empty, fill it with fountain water.
2. If the smaller jug is full, empty it to the fountain.
3. If there is water in the larger jug, pour it to the smaller jug until either the larger jug is empty or the smaller one is full.
4. Continue until there are n units of water in one jug.

Be Right and Be Fast

Discrete mathematicians and computer scientists focus on two aspects of algorithms in particular: correctness and speed. The former requires a formal proof that the algorithm delivers its output as promised, while the latter requires an analysis of the algorithm's efficiency. Although we have not rigorously proved that Algorithm 5.2 is correct—that we always end up at Step 4 with the desired amount of water—a deeper analysis of the billiard table method *does* establish correctness.

Condition (b) of the water measuring theorem provides some information about the speed of the algorithm by placing an upper bound on the number of pourings required. In fact, it is known that a careful choice of either Algorithm 5.2 or its reversed counterpart gives the smallest number of pourings for each desired amount of water. Figure 5.4 now makes it clear that Simon's demand for 4 gallons is the most difficult water measuring task involving 3- and 5-gallon jugs.

5.6 Arithmetic Arrays: Climb the Staircase

We introduce a new way to solve water measuring problems in which the abbreviated state sequence is read off directly from a staircase of numbers in a rectangular array. We first illustrate the method in some specific cases and then explain how it works in general.

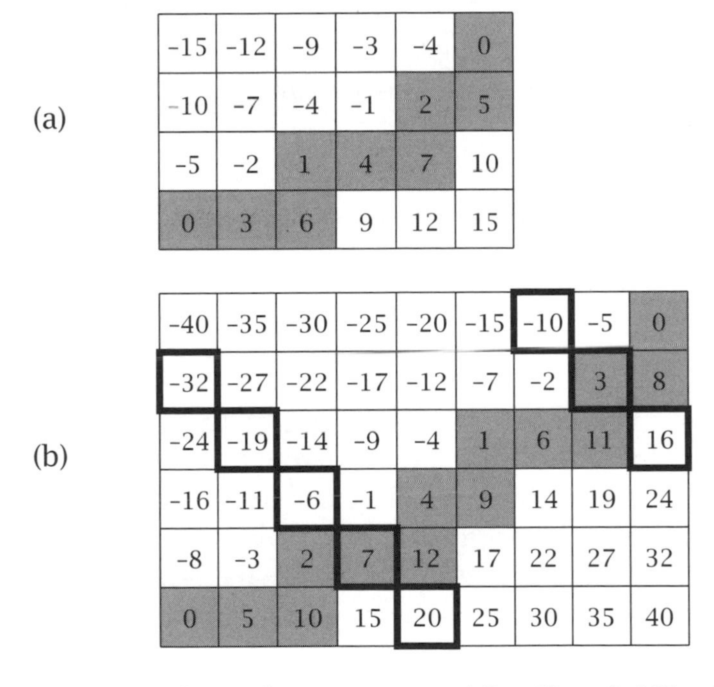

(a)

-15	-12	-9	-3	-4	0
-10	-7	-4	-1	2	5
-5	-2	1	4	7	10
0	3	6	9	12	15

(b)

-40	-35	-30	-25	-20	-15	-10	-5	0
-32	-27	-22	-17	-12	-7	-2	3	8
-24	-19	-14	-9	-4	1	6	11	16
-16	-11	-6	-1	4	9	14	19	24
-8	-3	2	7	12	17	22	27	32
0	5	10	15	20	25	30	35	40

Figure 5.6: The arithmetic arrays $A(3,-5)$ and $A(5,-8)$ contain abbreviated state sequences on the shaded staircases

Example 3. Figure 5.6(a) shows the array $A(3,-5)$. The rising shaded staircase steps through the abbreviated state sequence

$$0 \rightsquigarrow 3 \rightsquigarrow 6 \rightsquigarrow 1 \rightsquigarrow 4 \rightsquigarrow 7 \rightsquigarrow 2 \rightsquigarrow 5 \rightsquigarrow 0$$

for the Bruce Willis problem with 3- and 5-gallon jugs. (Each staircase sequence begins and ends with 0.) We have already observed that this abbreviated sequence solves all water measuring problems with 3- and 5-gallon jugs. Figure 5.6(b) shows the array $A(5,-8)$. The shaded staircase contains the abbreviated state sequence for 5- and 8-unit jugs:

$$0 \sim\sim 5 \sim\sim 10 \sim\sim 2 \sim\sim 7 \sim\sim 12 \sim\sim 4 \sim\sim \cdots \sim\sim 3 \sim\sim 8 \sim\sim 0.$$

We next show how to define the arrays and build the staircases.

The Arithmetic Array $A(v,-w)$

The *arithmetic array* $A(v,-w)$ contains the number

$$vx - wy$$

in column x and row y. There are $w+1$ columns and $v+1$ rows with column 0 on the left, and row 0 on the bottom. The entries increase by v across each row. Each row of $A(v,-w)$ contains an *arithmetic progression* with common difference v. Similarly, each column of $A(v,-w)$ contains an arithmetic progression with common difference $-w$ from bottom to top.

Arithmetic arrays also help us solve computer line drawing problems (Chapter 4) and explain properties of quadratic residues (Chapter 7).

How to Build a Staircase

The arithmetic array $A(v,-w)$ contains $v+w-1$ downward-sloping diagonals, each of which starts on the left or upper edge and ends on the right or lower edge. Two such diagonals are indicated in Figure 5.6(b). Each diagonal contains an arithmetic progression with common difference $v + w$. The first term of the progression is a nonpositive number and the last term is nonnegative. In each diagonal we shade the cell with smallest nonnegative number. The shaded cells form our staircase. Note that the numbers in the staircase must come from the list $0, 1, \ldots, v + w - 1$ since the arithmetic progressions on the diagonals have common difference $v + w$.

The abbreviated state sequence starts at 0 in the lower left corner with both jugs empty. A horizontal move on the staircase tells us to fill the smaller jug with fountain water, and a vertical move tells us to empty the larger jug. Other pourings are dictated by these conditions. (The water moves backward relative to the mnemonic in Figure 5.5.) The entry $vx - wy$ in column x and row y equals the total amount of water in the two jugs after the v-unit jug has been filled x times and the w-unit jug emptied y times.

The Arithmetic Array Theorem

The staircase method of solving water measuring problems assumes that each of the numbers $1, 2, \ldots, v + w - 1$ occurs on the staircase in the array $A(v, -w)$. Any omitted number would be an unmeasurable amount of water. The following result shows that no number is omitted.

Arithmetic array theorem. If v and w are co-prime, then the nonzero numbers in the arithmetic array $A(v, -w)$ are distinct. Also, 0 occurs twice—in the lower left and the upper right corners.

Let us first see how the theorem justifies our assumption. There are $v + w - 1$ positions on the staircase, excluding the two corners at the start and at the end. The numbers in these positions come from the list $1, 2, \ldots, v + w - 1$. The theorem assures us there are no repeats, and so each number in the list must occur exactly once on the staircase.

Now let us see why the arithmetic array theorem is true. It is easily verified for $v = 5$ and $w = 8$ by a direct examination of the 54 entries in $A(5, -8)$. But we want an algebraic confirmation that can be applied to any array. So suppose that the numbers in the distinct positions (x_1, y_1) and (x_2, y_2) are equal in $A(5, -8)$. Then

$$5x_1 - 8y_1 = 5x_2 - 8y_2.$$

We cannot have $x_1 = x_2$ because the numbers in each column form an arithmetic progression with difference -8. Let us say that $x_1 < x_2$. Rearrangement gives

$$\frac{5}{8} = \frac{y_2 - y_1}{x_2 - x_1}.$$

The fraction $5/8$ is in lowest terms. It follows that the denominator $x_2 - x_1$ is a positive multiple of 8. Because x_1 and x_2 are in the list 0, 1, ..., 8, the difference $x_2 - x_1$ can be a multiple of 8 only if $x_2 = 8$ and $x_1 = 0$. Similarly, $y_2 = 5$ and $y_1 = 0$. So the only two places with equal numbers are $(x_1, y_1) = (0, 0)$ and $(x_2, y_2) = (8, 5)$, giving the 0's in the lower left and the upper right corners of the array.

A similar argument applies to the general array $A(v, -w)$ provided v and w are co-prime.

The Co-primality Theorem (Again)

Let us revisit Bruce Willis and his 4 gallons of water. In carrying out the solution

$$\lfloor 0 \,|\, 0 \rfloor \sim\sim \lfloor 0 \,|\, 5 \rfloor \sim\sim \lfloor 3 \,|\, 2 \rfloor \sim\sim \lfloor 0 \,|\, 2 \rfloor \sim\sim \lfloor 2 \,|\, 0 \rfloor$$

$$\sim\sim \lfloor 2 \,|\, 5 \rfloor \sim\sim \lfloor 3 \,|\, 4 \rfloor \sim\sim \lfloor 0 \,|\, 4 \rfloor$$

from Example 1(b), the 3-gallon jug is emptied twice, and the 5-gallon jug is filled twice. Algebraically, this says that the integers $x = -2$ and $y = 2$ satisfy the equation

$$4 = 3x + 5y.$$

Another solution, $(x, y) = (3, -1)$, arises from the sequence of pourings $\lfloor 3 \mid 0 \rfloor \rightsquigarrow \lfloor 0 \mid 3 \rfloor \rightsquigarrow \lfloor 3 \mid 3 \rfloor \rightsquigarrow \cdots$ in Example 1(a). In general, Algorithm 5.2 yields integers x and y that satisfy the relation

$$n = vx + wy.$$

The most important case is $n = 1$, which we encountered in our discussion of Euclid's algorithm in Chapter 4.

Co-primality theorem. If v and w are co-prime, then there are integers x and y that satisfy $1 = vx + wy$.

5.7 Other Problems to Pour Over

Let us explore some variations on our basic water measuring theme.

Tartaglian Problems

The water measuring problem faced by Bruce Willis is a variant of a conundrum posed by Abbot Albert in the thirteenth century. The original problem has three jugs with capacities 3, 5, and 8 units. The 8-unit jug is full of water at the outset, and the object is to share the water equally between two people by getting 4 units of water in each of the two largest jugs. There is no fountain, and thus the water must be conserved at each pouring, a constraint that makes the problem somewhat more challenging. In general, there are three jugs with capacities v, w, and $v + w$ units, the largest of which is full of water initially. There is no fountain, and the goal is to get n units of water in one jug and $v + w - n$ units of water in another. Such problems are sometimes called *Tartaglian water measuring problems* because they appear in the work of the Italian mathematician Niccolò Tartaglia (1499–1557). A little experimentation reveals that the skew billiard table and Algorithm 5.2 solve Tartaglian problems, too, where the jug with capacity $v + w$ plays the role of the fountain. The site

```
maa.org/editorial/knot/water.html
```

by Alex Bogomolny includes an excellent interactive applet that allows you to adjust the jug capacities and experiment with Tartaglian water measuring problems of your own.

Tweedie also used billiard tables with more than four sides to solve water measuring problems in which the capacity of the largest jug is not the sum of the capacities of the two other jugs. See Problems 4–6 for some examples.

Three Jugs and a Fountain

A natural generalization of the water measuring problem starts with more jugs. We expect it to be easier to measure a desired amount of water with more jugs available. However, there are more choices to make at each stage, and finding an optimal sequence of pourings can be difficult.

Example 4. How can we measure 23 units of water from a fountain with 6-, 10-, and 15-unit jugs?

Here is one solution. First, fill the 15-unit jug, and then use the two other jugs to measure 8 units. The latter step is essentially a doubling of the Bruce Willis problem of measuring 4 units with 3- and 5-unit jugs. If we let $\lfloor a \mid b \mid c \rfloor$ denote the state with a, b, and c units of water, respectively, in the 6-, 10-, and 15-unit jugs, then this solution is

$$\lfloor 0 \mid 0 \mid 0 \rfloor \rightsquigarrow \lfloor 0 \mid 0 \mid 15 \rfloor \rightsquigarrow \lfloor 0 \mid 10 \mid 15 \rfloor$$

$$\rightsquigarrow \lfloor 6 \mid 4 \mid 15 \rfloor \rightsquigarrow \lfloor 0 \mid 4 \mid 15 \rfloor \rightsquigarrow \lfloor 4 \mid 0 \mid 15 \rfloor$$

$$\rightsquigarrow \lfloor 4 \mid 10 \mid 15 \rfloor \rightsquigarrow \lfloor 6 \mid 8 \mid 15 \rfloor \rightsquigarrow \lfloor 0 \mid 8 \mid 15 \rfloor.$$

Another solution is

$$\lfloor 0 \mid 0 \mid 0 \rfloor \rightsquigarrow \lfloor 0 \mid 0 \mid 15 \rfloor \rightsquigarrow \lfloor 0 \mid 10 \mid 5 \rfloor$$

$$\rightsquigarrow \lfloor 0 \mid 0 \mid 5 \rfloor \rightsquigarrow \lfloor 6 \mid 0 \mid 5 \rfloor \rightsquigarrow \lfloor 0 \mid 0 \mid 11 \rfloor$$

$$\rightsquigarrow \lfloor 6 \mid 0 \mid 11 \rfloor \rightsquigarrow \lfloor 0 \mid 6 \mid 11 \rfloor \rightsquigarrow \lfloor 6 \mid 6 \mid 11 \rfloor.$$

Each of the two preceding solutions uses eight pourings. Is there a solution with fewer pourings? In the two-jug problem the circular arrangement of states (Figure 5.4) makes it easy to find the shortest route to measure a specified amount of water. However, the many states in a three-jug problem are connected by a complicated network of pourings, making it difficult to extract the shortest route to measure a given amount of water. For instance, Figure 5.7 shows

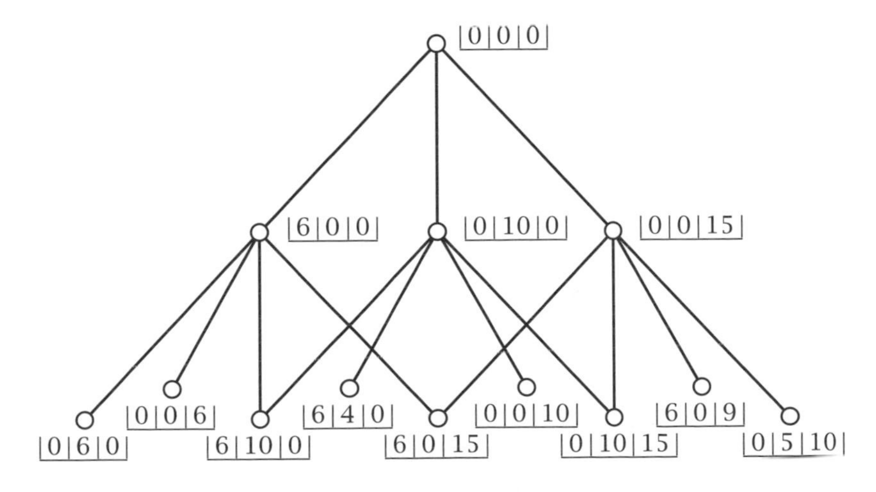

Figure 5.7: Some states for a three-jug problem

the thirteen possible states that arise with at most two pourings in the situation of Example 4. It is clear that the entire state diagram will be unwieldy.

Many Jugs

What amounts of water can we measure if we have j jugs with capacities v_1, v_2, ..., v_j at a fountain? We continue to assume that the jug capacities are co-prime:

$$\gcd(v_1, v_2, \ldots, v_j) = 1.$$

The problem is wholly uninteresting with one jug. The water measuring theorem treats the two-jug problem. The following theorem covers the cases with more than two jugs.

General water measuring theorem. If we are at a fountain with j unmarked jugs ($j \geq 2$) with co-prime capacities v_1, v_2, ..., v_j, then we can measure n units of water for $n = 0, 1, \ldots, v_1 + v_2 + \cdots + v_j$.

To prove the theorem, we might begin by finding integers $x_1, x_2, \ldots, x_j$ that satisfy

$$n = v_1x_1 + v_2x_2 + \cdots + v_jx_j. \tag{5.1}$$

The positive and negative values of x_1, x_2, ..., x_j give hints about the number of times the jugs should be filled or emptied to measure n units of water. For instance, the solution $(x, y, z) = (-2, 2, 1)$ to the equation

$$23 = 6x + 10y + 15z$$

corresponds to the first method of measuring 23 units in Example 4 with the smallest jug emptied twice, the middle jug filled twice, and the largest jug filled once. A careful analysis is needed to turn this idea into a rigorous general argument. Moreover, equation (5.1) has many solutions in general, and it is not clear how to locate the ones that yield minimal sequences of pourings. Problem 14 gives more information.

Interstate Travel

Our water measuring problems have all started with both jugs empty. We can also ask for the smallest number of pourings needed to pass from a given initial state to a target state. A circle of states like the one in Figure 5.4 is inadequate for these purposes because some shortcuts are absent. For instance, with 3- and 5-unit jugs, Figure 5.4 shows that we can pass from ⌊3|2⌋ to ⌊3|0⌋ in three pourings. But it is clear that just one pouring is enough; simply dump the 2 units of water from the 5-unit jug.

Solving these state-to-state problems in the minimum number of moves is facilitated by a complete *state diagram*

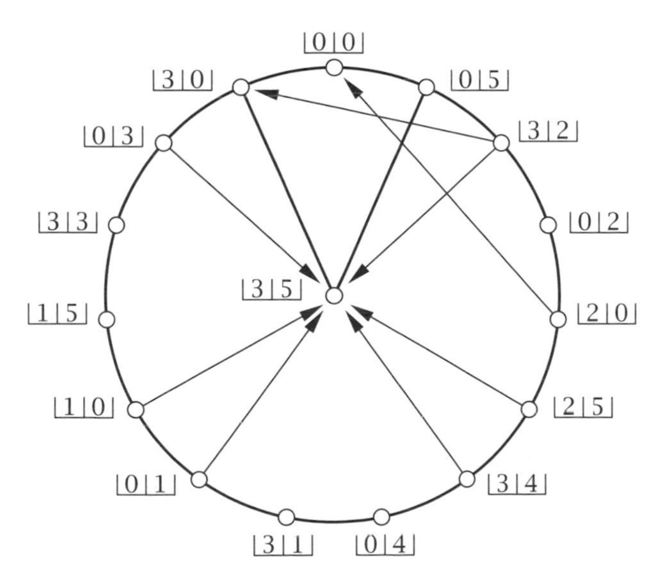

Figure 5.8: A more complete state diagram for 3- and 5-unit jugs

that shows all direct connections between states. Figure 5.8 includes the direct connection from $\lfloor 3 \mid 2 \rfloor$ to $\lfloor 3 \mid 0 \rfloor$, where the arrow indicates that the process cannot be reversed. There are many similar moves not indicated in the figure. Note that the state diagram also includes the state $\lfloor 3 \mid 5 \rfloor$ with both jugs full.

5.8 Number Theory and Fermat's Congruence

The French mathematician Pierre de Fermat (1601–1665) noticed a divisibility pattern involving numbers raised to powers. The numbers

$$1^3 - 1, \quad 2^3 - 2, \quad 3^3 - 3, \quad 4^3 - 4, \quad 5^3 - 5, \quad \ldots$$

are all divisible by 3, the numbers

$$1^5 - 1, \quad 2^5 - 2, \quad 3^5 - 3, \quad 4^5 - 4, \quad 5^5 - 5, \quad \ldots$$

are all divisible by 5, and the numbers

$$1^7 - 1, \quad 2^7 - 2, \quad 3^7 - 3, \quad 4^7 - 4, \quad 5^7 - 5, \quad \ldots$$

are all divisible by 7. Alas, the pattern breaks down for the exponent 9 since $2^9 - 2 = 510$, which is not divisible by 9. However, the pattern works for exponents 11 and 13. Fermat discovered that the pattern works whenever the exponent is a prime. A *prime* is a positive number p with exactly two divisors, namely, 1 and p itself. The first few primes are 2, 3, 5, 7, 11, and 13.

Fermat's theorem. If the exponent p is a prime, then the numbers

$$1^p - 1, \quad 2^p - 2, \quad 3^p - 3, \quad 4^p - 4, \quad 5^p - 5, \quad \ldots$$

are divisible by p.

Fermat seldom shared his proofs or methods, preferring to leave the demonstrations of his discoveries as challenges for his contemporaries. In this section, we use our arithmetic arrays to explain why Fermat's result is true. Our argument is certainly not the one he had in mind for his theorem. Perhaps he relied on the binomial theorem, as outlined in Problem 12.

Modular Arithmetic

Fermat's result is best understood in the language of modular arithmetic. The number a is *congruent* to b modulo m, written

$$a \equiv b \pmod{m},$$

provided $a - b$ is divisible by m. The *modulus* m is a positive integer, but a and b may be positive, negative, or zero. For every integer a, there is a unique integer r among 0, 1, 2, $\ldots, m-1$ such that $a \equiv r \pmod{m}$. If a is nonnegative, then r is the remainder when a is divided by m.

Example 5. (a) We have $510 \equiv 6 \pmod{9}$ since $510 - 6 = 504$, which is divisible by 9.

(b) Even numbers are congruent to 0 modulo 2, and odd numbers are congruent to 1 modulo 2.

(c) If today is a Thursday, then in 100 days it will be a Saturday since $100 \equiv 2 \pmod{7}$.

(d) The positive integers a and b satisfy $a \equiv b \pmod{10}$ provided they have the same last digit.

The congruence symbol resembles an equals sign, a felicitous choice of notation since congruences enjoy many of the same properties as equalities. For instance, transitivity holds: If $a \equiv b \pmod{m}$ and $b \equiv c \pmod{m}$, then $a \equiv c \pmod{m}$. Also, if $a \equiv a' \pmod{m}$ and $b \equiv b' \pmod{m}$, then $a + b \equiv a' + b' \pmod{m}$ and $ab \equiv a'b' \pmod{m}$.

Example 6. There is a straightforward method to simplify $2^9 - 2$ modulo 9, namely, raise 2 to the ninth power, subtract 2, divide by 9, and take the remainder. A less tedious method uses properties of modular arithmetic to streamline the calculations. Because $2^3 \equiv 8 \equiv -1 \pmod{9}$, we have

$$2^9 - 2 \equiv (2^3)^3 - 2 \equiv (-1)^3 - 2 \equiv -3 \equiv 6 \pmod{9}.$$

Our proof of Fermat's congruence relies on a type of cancellation in modular arithmetic. If the modulus m is a prime and $y \not\equiv 0 \pmod{m}$, then we may cancel the factor y from both sides of the congruence $ay \equiv by \pmod{m}$ to conclude that $a \equiv b \pmod{m}$. Note that if the modulus m is not a prime, then such cancellation might be invalid. For instance, $3 \times 2 \equiv 0 \times 2 \pmod{6}$, but $3 \not\equiv 0 \pmod{6}$.

Fermat's Congruence and Arithmetic Arrays

In the notation of modular arithmetic, Fermat's result takes the following form.

Fermat's congruence. If v is an integer and p is a prime, then

$$v^p \equiv v \pmod{p}.$$

We now explain why Fermat's congruence is true with the help of the same arithmetic arrays we used to solve water measuring problems. We start with two observations that simplify our discussion. Because every integer is congruent to one of the numbers $0, 1, \ldots, p-1$ modulo p, we only need to prove Fermat's congruence for $v = 0, 1, \ldots, p-1$. Also, the congruence clearly holds for $v = 0$. So we can restrict our attention to $v = 1, 2, \ldots, p-1$.

Our argument hinges on properties of the arithmetic array $A(v, -p)$ with entry

$$vx - py$$

in the intersection of column x and row y. We focus on the positions of the numbers 1, 2, ..., $p - 1$ in the array. Each of these numbers must occur *somewhere* in the array—in fact, on the staircase used for the water measuring problem. Figure 5.9 shows the case $v = 4$ and $p = 7$. Let n occur in

-28	-24	-20	-16	-12	-8	-4	0
-21	-17	-13	-9	-5	-1	3	7
-14	-10	-6	-2	2	6	10	14
-7	-3	1	5	9	13	17	21
0	4	8	12	16	20	24	28

Figure 5.9: The arithmetic array $A(v, -p)$ for $v = 4$ and $p = 7$

column x_n and row y_n for $n = 1, 2, \ldots, p - 1$. Then the columns $x_1, x_2 \ldots, x_{p-1}$ are chosen from $1, 2, \ldots, p - 1$. We claim that the numbers in the two lists

$$x_1, x_2, \ldots, x_{p-1} \qquad \text{and} \qquad vx_1, vx_2, \ldots, vx_{p-1}$$

are both rearrangements of $1, 2, \ldots, p - 1$ modulo p. First, no two of the $p - 1$ consecutive numbers $1, 2, \ldots, p - 1$ can occur in the same column of $A(v, -p)$ since each column is an arithmetic progression with difference $-p$. This implies that the list $x_1, x_2, \ldots, x_{p-1}$ is a rearrangement of 1, 2, ..., $p - 1$. Second, every number in column x of $A(v, -p)$ is congruent to vx modulo p since

$$n = vx_n - py_n \equiv vx_n \pmod{p}.$$

It follows that the list $vx_1, vx_2, \ldots, vx_{p-1}$ is another rearrangement of $1, 2, \ldots p - 1$ modulo p. The products of the numbers in our two lists must be equal to one another modulo p:

$$(vx_1)(vx_2)\cdots(vx_{p-1}) \equiv x_1x_2\cdots x_{p-1} \pmod p$$

Thus

$$v^{p-1}x_1x_2\cdots x_{p-1} \equiv x_1x_2\cdots x_{p-1} \pmod p.$$

Cancel the product $x_1x_2\cdots x_{p-1}$ from both sides to arrive at the congruence

$$v^{p-1} \equiv 1 \pmod p.$$

Finally, multiply both sides by v to get Fermat's congruence, $v^p \equiv v \pmod p$.

The co-primality theorem and Fermat's congruence are used throughout discrete mathematics and are cornerstones of modern cryptography. The fast, secure transactions of debit and ATM cards rely on both of these results, for example, as do some digital signature protocols. A taste of this type of application is given in Section 7.6.

5.9 Notes and References

The skew billiard table was introduced by Tweedie [14]. The states on the edges of the billiard table are an instance of trilinear coordinates. See the geometry book [5] for more background. Bailey [1] solves water measuring problems by using modular arithmetic to generate our abbreviated state sequences directly.

Algorithms for Tartaglian and state-to-state water measuring problems are examined in [2] and [8]. Problems with more jugs are studied in [10]. The general water measuring theorem for j jugs is a consequence of a result of Boldi, Santini, and Vigna [4].

For background on modular arithmetic, Fermat's congruence, and many other topics in number theory see [12]. Many books explain the role of modular arithmetic in cryptography, e.g., [3] and [13]. Methods to solve equation (5.1) are discussed in Kertzner [7]. Problem 7 is adapted from [6].

1. Bailey, H., Fetching water with least residues, *College Mathematics Journal* **39** (2008), 304–306.
2. Bellman, R., Cooke, K. L., and Lockett, J. A., *Algorithms, Graphs, and Computers.* Academic Press, New York. 1970.
3. Beutelspacher, Albert, *Cryptology.* Mathematical Association of America, Washington, DC, 1994.
4. Boldi, P., Santini, M., and Vigna, S., Measuring with jugs, *Theoretical Computer Science* **282** (2002), 259–270.
5. Coxeter, H. S. M., and Greitzer, S. L., *Geometry Revisited.* Mathematical Association of America, Washington, DC, 1967.
6. Hess, R. I., Bonus problem, *The Bent of Tau Beta Pi* **96** (2005), No. 3, 53.
7. Kertzner, S., The linear diophantine equation, *American Mathematical Monthly* **88** (1981), 200–203.
8. McDiarmid, C. J. H., and Ramírez Alfonsín, J. L., Sharing jugs of wine, *Discrete Mathematics* **125** (1994), 279–287.
9. O'Bierne, T. H., *Puzzles and Paradoxes.* Oxford University Press, New York, 1965.
10. Pfaff, T. J., and Tran, M. M., The generalized jug problem, *Journal of Recreational Mathematics* **31** (2002–03), 100–103.
11. Sawyer, W. W., On a well-known puzzle, *Scripta Mathematica* **16** (1950), 107–110.
12. Stillwell, John, *Elements of Number Theory.* Springer, New York, 2003.
13. Stinson, Douglas R., *Cryptology: Theory and Practice,* 2nd ed. Chapman Hall/CRC, Boca Raton, Florida, 2002.
14. Tweedie, M. C. K., A graphical method of solving Tartaglian measuring puzzles, *Mathematical Gazette* **23** (1939), 278–282.

5.10 Problems

1. Find a fast way to measure 1 gallon of water with cylindrical 3- and 5-gallon jugs if the tilting illustrated in Figure 5.1 is allowed.

2. Solve the water measuring problem in Example 2 with fewer than 19 pourings.

3. We have unmarked jugs with capacities 7 and 10 units at a fountain. The smaller jug is empty, and the larger jug has 2 units of water in it. Measure 1 unit of water with as few pourings as possible.

4. We have an 8-unit jug filled with water and empty 3- and 7-unit jugs. The jugs are unmarked, and there is no fountain. Each state is of the form $\lfloor a \,|\, b \,|\, c \rfloor$ with $a + b + c = 8$.

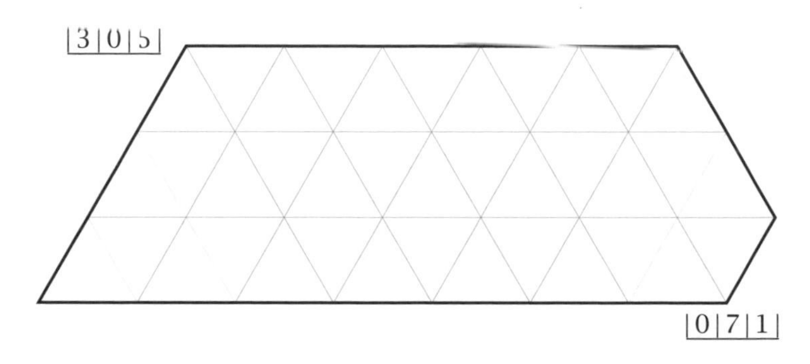

Figure 5.10: Billiard table for 3-, 7-, and 8-unit jugs

   (a) Complete the labels on the billiard table in Figure 5.10.

   (b) Explain why the billiard table is not a parallelogram for this problem.

   (c) Give instructions to arrive at the state $\lfloor 0 \,|\, 4 \,|\, 4 \rfloor$.

5. We have 6-, 7-, and 8-unit jugs and no fountain. There are 10 units of water distributed among the jugs in some manner. Each state is of the form $\lfloor a \,|\, b \,|\, c \rfloor$, where $a + b + c = 10$.

   (a) Create a suitable billiard table. Hint: Two of the six corner states are $\lfloor 0 \,|\, 7 \,|\, 3 \rfloor$ and $\lfloor 2 \,|\, 0 \,|\, 8 \rfloor$.

   (b) Can two people share the water equally if the initial state is $\lfloor 0 \,|\, 7 \,|\, 3 \rfloor$? If so, how?

6. We have unmarked 5-, 6- and 10-unit jugs containing 0, 6, and 6 units of water, respectively. Show how to arrive at a state with 3 units of water in one jug, 4 units in another, and 5 units in the remaining jug. There is no fountain.

7. We have a 15-unit jug filled with water and want to distribute 5, 8, and 2 units, respectively, to 6-, 10-, and q-unit jugs. Here q is not known, but $q \geq 2$, of course. Find a sequence of pourings.

8. Add a few lines to Algorithm 5.2 so that it also computes integers x and y that satisfy $n = vx + wy$.

9. Draw a complete state diagram for the Tartaglian water measuring problem with 3-, 5-, and 8-unit jugs. Use labels of the form $\lfloor a \,|\, b \,|\, c \rfloor$, where $a + b + c = 8$.

10. Extend Figure 5.7 to include all states that can be achieved with at most three pourings.

11. Let v and w be co-prime and let n be an integer. Show that there is an integer x satisfying $vx \equiv n \pmod{w}$.

12. This problem outlines another proof of Fermat's congruence. Recall that the choose function $\binom{p}{k}$ is given by

$$\binom{p}{k} = \frac{p!}{k!\,(p-k)!}.$$

The binomial theorem says that

$$(x+1)^p = x^p + \binom{p}{1}x^{p-1} + \binom{p}{2}x^{p-2} + \cdots + \binom{p}{p-1}x + 1.$$

Let p be a prime.

(a) Explain why $\binom{p}{k}$ is divisible by p for $k = 1, 2, \ldots, p-1$.

(b) Deduce that $(x+1)^p \equiv x^p + 1 \pmod{p}$.

(c) Use part (b) with $x = 1$ to show that $2^p \equiv 2 \pmod{p}$

(d) Show that $3^p \equiv 3 \pmod{p}$.

(e) Show that Fermat's congruence $v^p \equiv v \pmod{p}$ is valid for each integer v.

-40	-35	-30	-25	-20	-15	-10	-5	0
-32	-27	-22	-17	-12	-7	-2	3	8
-24	-19	-14	-9	-4	1	6	11	16
-16	-11	-6	-1	4	9	14	19	24
-8	-3	2	7	12	17	22	27	32
0	5	10	15	20	25	30	35	40

Figure 5.11: Another staircase in $A(5, -8)$

13. Figure 5.11 shows a new type of staircase in the arithmetic array $A(5, -8)$.

    (a) How is the new staircase defined in the array $A(v, -w)$?

    (b) Explain how the new staircase helps solve water measuring problems.

14. Let the jug capacities v_1, v_2, ..., v_j be co-prime, that is,

$$\gcd(v_1, v_2, \ldots, v_j) = 1.$$

Let v_j be the largest capacity, and suppose that $0 \le n \le v_1 + v_2 + \cdots + v_j$. Boldi et al. proved that it is possible to place $\hat{n}$ units of water from a fountain in the largest jug for any amount $\hat{n}$ satisfying $0 \le \hat{n} \le v_j$. Use this result to show that we can measure n units of water from a fountain with the j jugs. This is our general water measuring theorem. Hint: First suppose that $n - \hat{n}$ is positive and divisible by $\gcd(v_1, v_2, \ldots, v_{j-1})$.

That which the fountain sends forth
returns again to the fountain.
HENRY WADSWORTH LONGFELLOW

6

From Stamps to Sylver Coins

The great ideas in modern mathematics have their origin in observation.
J. J. SYLVESTER

6.1 Sylvester's Stamps

Question. We have a large supply of 5- and 8-cent stamps. Can we make exact postage for a 27-cent postcard?

A bit of trial and error should convince you the answer is *no.* A methodical approach involves some algebra. The question asks whether there is a solution (x, y) to the equation

$$27 = 5x + 8y,$$

where, of course,

$$x = \text{the number of 5-cent stamps}$$
$$\text{and} \quad y = \text{the number of 8-cent stamps.}$$

The numbers x and y must be nonnegative integers. When we try $x = 0, 1, \ldots, 5$ in succession, none of the corresponding values of y is a nonnegative integer. There is no need to try values of x greater than 5, and we have confirmed our answer to the postage question.

More generally, to achieve exactly n cents postage with 5- and 8-cent stamps, we must find a pair of nonnegative integers (x, y) satisfying

$$n = 5x + 8y.$$

Some amounts less than 27 cents are attainable (e.g., 5, 8, 10, and 23), while others are not (e.g., 1, 4, 11, and 22). A detailed analysis shows that 14 amounts cannot be attained exactly, namely,

$$1,\ 2,\ 3,\ 4,\ 6,\ 7,\ 9,\ 11,\ 12,\ 14,\ 17,\ 19,\ 22,\ 27.$$

No amount greater than 27 is attainable, as we will see later.

The influential English mathematician James Joseph Sylvester (1814–1897) contemplated the general question for stamps with denominations v and w. He examined nonnegative integer solutions (x, y) to the equation

$$n = vx + wy,$$

where v and w are given, and n is a desired postage. When the equation has a solution (x, y), we say that n is *attainable* by v and w, or just *attainable* if the values of v and w are clear from context. When there is no solution, n is *unattainable.*

Additive Number Theory

As you probably guessed, we are not interested in actual postal applications. The stamps merely provide a scenario for a fundamental question in *additive number theory,* the branch of pure mathematics that studies ways to represent a number as the sum of other numbers of specified types. Some examples capture the spirit of the subject.

Example 1. (a) A solution (x, y) to the stamp equation

$$n = vx + wy$$

produces a representation of n as the sum of nonnegative multiples of the given numbers v and w.

(b) The dollar-changing problem asks for the number of ways to make change for D dollars from a supply of quarters, dimes, and nickels. (See Chapter 2.) This is the same as counting the ways to represent $100D$ as the sum of nonnegative multiples of 25, 10, and 5.

(c) When are the three sides of a right triangle integers? In algebraic terms, when is the square of a number the sum of two other squares? Familiar examples include $5^2 = 3^2 + 4^2$ and $13^2 = 5^2 + 12^2$. We want to construct and classify all *Pythagorean triples*—triples of positive integers (a, b, c) satisfying

$$c^2 = a^2 + b^2.$$

(d) The *four squares theorem* asserts that every positive integer can be expressed as the sum of at most four squares, e.g.,

$$43 = 25 + 9 + 9 \qquad \text{and} \qquad 666 = 625 + 36 + 4 + 1.$$

This theorem was first proved in 1770 by Joseph Louis Lagrange (1736–1813), whose penetrating insights influenced many areas of mathematics. The four squares theorem epitomizes the austere elegance for which additive number theory is known. Its brief, readily grasped statement belies the subtlety of its proof.

(e) The *partition problem* asks for the number of ways to represent each positive integer as a sum of one or more positive integers. For instance, the five partitions of 4 are

$$4, \quad 3 + 1, \quad 2 + 2, \quad 2 + 1 + 1, \quad 1 + 1 + 1 + 1.$$

The Stamp Problem

While investigating complex aspects of the partition problem, Sylvester found an elementary method to count some special solutions of the stamp equation. His discovery was buried in a comment within a lengthy appendix to a formidable 56-page article[1] published in 1882. To advertise his elementary approach more widely, he later posed a challenge to readers of a popular mathematics journal.

Stamp problem. If our stamps have co-prime denominations v and w, how many unattainable postages are less than vw?

All amounts are attainable if either $v = 1$ or $w = 1$, an uninteresting situation we will usually exclude. Recall that two positive numbers are *co-prime* provided their greatest common divisor is 1. For instance, 5 and 8 are co-prime, but 8 and 12 are not. Sylvester's restriction to co-prime stamp denominations is entirely natural since we may always divide by the greatest common divisor to produce a co-prime situation. See Problem 13 for details.

On the other hand, Sylvester's restriction to amounts less than vw is an unfortunate artifact of the elementary solution he had in mind. We will see (and Sylvester surely knew) that there are no unattainable numbers greater than vw.

Our first goal in this chapter is to solve the stamp problem. Our solution is not the most direct, but it does provide substantial insight and allows us to answer a sharper version of Sylvester's original question. Our approach exploits symmetry in rectangular arrays of numbers. An important result in number theory, the Chinese Remainder Theorem, is a pleasant by-product. We also look at what happens when more than two denominations are available and explore a mind-boggling game known as sylver coinage.

---

[1]Problem 9 quotes the relevant passage and offers a sample of Sylvester's inimitable writing style.

6.2 Addition Tables and Symmetry

Let us investigate the stamp problem for the case $v = 5$ and $w = 8$ in detail. Table 6.1 lists the representations of the 26 attainable numbers less than 40. Our inclusion of 0 as an attainable number dismays the post office but makes perfect mathematical sense.

Table 6.1: Attainable numbers less than 40 for $v = 5$ and $w = 8$

$$
\begin{array}{rcl@{\qquad}rcl}
0 &=& 5\times 0 + 8\times 0 & 26 &=& 5\times 2 + 8\times 2 \\
5 &=& 5\times 1 + 8\times 0 & 28 &=& 5\times 4 + 8\times 1 \\
8 &=& 5\times 0 + 8\times 1 & 29 &=& 5\times 1 + 8\times 3 \\
10 &=& 5\times 2 + 8\times 0 & 30 &=& 5\times 6 + 8\times 0 \\
13 &=& 5\times 1 + 8\times 1 & 31 &=& 5\times 3 + 8\times 2 \\
15 &=& 5\times 3 + 8\times 0 & 32 &=& 5\times 0 + 8\times 4 \\
16 &=& 5\times 0 + 8\times 2 & 33 &=& 5\times 5 + 8\times 1 \\
18 &=& 5\times 2 + 8\times 1 & 34 &=& 5\times 2 + 8\times 3 \\
20 &=& 5\times 4 + 8\times 0 & 35 &=& 5\times 7 + 8\times 0 \\
21 &=& 5\times 1 + 8\times 2 & 36 &=& 5\times 4 + 8\times 2 \\
23 &=& 5\times 3 + 8\times 1 & 37 &=& 5\times 1 + 8\times 4 \\
24 &=& 5\times 0 + 8\times 3 & 38 &=& 5\times 6 + 8\times 1 \\
25 &=& 5\times 5 + 8\times 0 & 39 &=& 5\times 3 + 8\times 3
\end{array}
$$

The entries in the table can be generated one at a time by trial and error. The attainable numbers pop up more frequently as we proceed, and once the five consecutive attainable numbers 28, 29, 30, 31, and 32 occur, multiples of 5 can be added to represent all larger numbers. This confirms that 27 is the largest unattainable number. The 14 unattainable numbers less than 40 listed earlier are absent from the table.

We will continue to use an indirect counting method to solve the stamp problem. Even though the problem deals with unattainable numbers, it is more convenient to generate and count the attainable numbers and then use the basic equation

$$\begin{pmatrix} \text{the number of} \\ \text{unattainable numbers} \\ \text{less than } vw \end{pmatrix} = vw - \begin{pmatrix} \text{the number of} \\ \text{attainable numbers} \\ \text{less than } vw \end{pmatrix}.$$

Addition Tables

If you constructed Table 6.1 by hand, you likely discovered the following time-saver.

Fact. If n_1 and n_2 are attainable by v and w, then so is the sum $n_1 + n_2$.

This fact underlies a method that avoids the tedious trial and error process of Table 6.1, relying instead on a variant of the addition tables memorized by schoolchildren. Table 6.2 starts with the multiples of 5 across the bottom row and the multiples of 8 up the left column. The sums in the body of the table are attainable by 5 and 8. Extended indefinitely, the table includes all attainable numbers in all

Table 6.2: An addition table for multiples of 5 and 8

⋮	⋮	⋮	⋮	⋮	⋮	⋮	⋮	⋮	⋮	
40	40	45	50	55	60	65	70	75	80	⋯
32	32	37	42	47	52	57	62	67	72	⋯
24	24	29	34	39	44	49	54	59	64	⋯
16	16	21	26	31	36	41	46	51	56	⋯
8	8	13	18	23	28	33	38	43	48	⋯
0	0	5	10	15	20	25	30	35	40	⋯
+	0	5	10	15	20	25	30	35	40	⋯

possible ways. The 26 attainable numbers less than 40 are clustered in the shaded region in the lower left corner.

Symmetry and a Spin

Figure 6.1 exploits symmetry in a rectangular array to determine the number of attainable numbers without resorting to a direct count. The array contains the essential part of the addition table in Figure 6.2. The attainable sum

$$5x + 8y$$

occurs in column x and row y. (Columns are labeled $x = 0$, 1, ..., 8 from left to right, and rows are labeled $y = 0$, 1, ..., 5 from bottom to top.) The dark gray region contains numbers less than 40, while the light gray region contains numbers greater than 40.

attainable numbers less than 40

40	45	50	55	60	65	70	75	80
32	37	42	47	52	(57)	62	67	72
24	29	34	39	44	40	54	59	64
16	21	26	31	36	41	46	51	56
8	13	18	(23)	28	33	38	43	48
0	5	10	15	20	25	30	35	40

Figure 6.1: Pairs of attainable numbers sum to 80

To see why the two shaded regions must have the same number of cells, we spin the table 180° about its center point (Figure 6.2). The spin interchanges the two shaded regions, which means they must have the same number of cells. More precisely, the spin interchanges the cells (x, y)

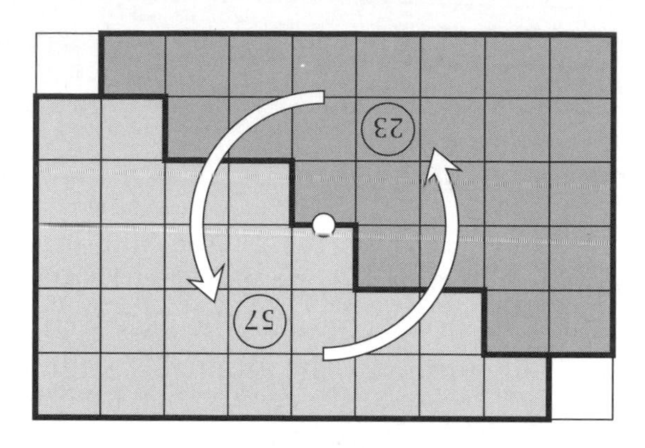

Figure 6.2: A 180° spin interchanges the shaded regions

and $(8-x, 5-y)$. The numbers in these two cells sum to 80 because

$$\left(5x + 8y\right) + \left(5(8-x) + 8(5-y)\right) = 80.$$

Figures 6.1 and 6.2 spotlight the sum $23+57 = 80$. It follows that each attainable number less than 40 has a mate greater than 40 in the array. Since the total number of cells in the array is $(5+1)(8+1)$, we know that

$$\begin{array}{c}\text{the number of attainable} \\ \text{numbers less than 40}\end{array} = \frac{(5+1)(8+1)-2}{2} = 26.$$

We subtracted 2 in the numerator to eliminate the two unshaded corners of the array, both of which contain the number 40. Once again we find that there are $40 - 26 = 14$ numbers less than 40 unattainable by $v = 5$ and $w = 8$.

6.3 Arithmetic Arrays and Sylvester's Formula

The ideas in the previous section work whenever the stamp denominations v and w are co-prime. We use the *arithmetic array* $A(v,w)$ with entry

$$vx + wy$$

in column x and row y. The array has $(v+1)(w+1)$ entries, and the number vw occurs in the upper left and lower right corners. A 180° spin of the array about its center pairs cell (x,y) with cell $(w-x, v-y)$. The numbers in these cells sum to $2vw$ since

$$\left(vw + wy\right) + \left(v(w-x) + w(v-y)\right) = 2vw.$$

It follows that each attainable number less than vw has a mate greater than vw in the array, and

$$\begin{matrix}\text{the number of attainable} \\ \text{numbers less than } vw\end{matrix} = \frac{(v+1)(w+1)-2}{2}.$$

Therefore, the number of unattainable numbers less than vw is

$$vw - \frac{(v+1)(w+1)-2}{2} = \frac{(v-1)(w-1)}{2}.$$

We have solved Sylvester's stamp problem.

Sylvester's formula. If v and w are co-prime, then the number of unattainable numbers less than vw is

$$\frac{(v-1)(w-1)}{2}.$$

Our visual demonstration for Sylvester's formula is convincing and appealing. But our argument rests on a tacit assumption.

Assumption 1. If v and w are co-prime, then none of the attainable numbers in the shaded regions of the arithmetic array $A(v,w)$ occurs more than once.

If the assumption were false, our enumeration of attainable and unattainable numbers would be wrong. By the way, the assumption *is* false when v and w are not co-prime (construct the array $A(4,6)$ to see what happens), but the stamp formula excludes this situation by hypothesis. We will justify assumption 1 later.

A Sharp Solution

In 1884, C. W. Curran Sharp used polynomials to solve Sylvester's problem. Here is how Sharp's solution works when $v = 5$ and $w = 8$. To count the attainable numbers less than 40, look at the polynomial product

$$P(z) = (1 + z^5 + z^{10} + \cdots + z^{40})\,(1 + z^8 + z^{16} + \cdots + z^{40})\,.$$

Some algebra produces the expansion

$$P(z) = 1 + z^5 + z^8 + \cdots + 2z^{40} + \cdots + z^{72} + z^{75} + z^{80}.$$

We can glean greater insight from the general form and properties of the expansion than from a complete listing of the individual terms. A typical term in the expansion of $P(z)$ is of the form z^n, where the exponent $n = 5x + 8y$ is an attainable number with x chosen from 0, 1, ..., 8 and y from 0, 1, ..., 5. Each attainable number from 0 to 39 occurs as an exponent. If z^n occurs with exponent $n = 5x + 8y$, then z^{80-n} also occurs since

$$80 - n = 5(8 - x) + 8(5 - y).$$

This means that the terms in the expansion of $P(z)$ occur in a symmetric pattern. The central term is $2z^{40}$.

Let us evaluate the polynomial $P(z)$ at $z = 1$ in two different ways. On the one hand,

$$\begin{aligned} P(1) &= (1 + 1^5 + 1^{10} + \cdots + 1^{40})\,(1 + 1^8 + 1^{16} + \cdots + 1^{40}) \\ &= 9 \times 6. \end{aligned}$$

On the other hand, each attainable number n from 0 to 39 gives rise to the terms z^n and z^{80-n}, which together contribute 2 to $P(1)$. The central term $2z^{40}$ also contributes 2 to $P(1)$, and so

$$P(1) = 2 \times \left(\begin{array}{c} \text{the number of attainable} \\ \text{numbers less than 40} \end{array} \right) + 2.$$

Our two expressions for $P(1)$ must be equal to each other, of course, and we conclude that

$$\begin{array}{c} \text{the number of attainable} \\ \text{numbers less than 40} \end{array} = \frac{9 \times 6 - 2}{2} = 26,$$

as we saw before. Sharp solved Sylvester's problem by applying the same argument to the product

$$(1 + z^v + z^{2v} + \cdots + z^{wv})(1 + z^w + z^{2w} + \cdots + z^{vw}).$$

You no doubt noticed the similarities between Sharp's approach and our use of arithmetic arrays. Table 6.3 makes the connection explicit. Each monomial z^n in Sharp's polynomial product corresponds to an attainable number n in our arithmetic array $A(5,8)$.

Table 6.3: A multiplication table

z^{40}	z^{40}	z^{45}	z^{50}	z^{55}	z^{60}	z^{65}	z^{70}	z^{75}	z^{80}
z^{32}	z^{32}	z^{37}	z^{42}	z^{47}	z^{52}	z^{57}	z^{62}	z^{67}	z^{72}
z^{24}	z^{24}	z^{29}	z^{34}	z^{39}	z^{44}	z^{49}	z^{54}	z^{59}	z^{64}
z^{16}	z^{16}	z^{21}	z^{26}	z^{31}	z^{36}	z^{41}	z^{46}	z^{51}	z^{56}
z^{8}	z^{8}	z^{13}	z^{18}	z^{23}	z^{28}	z^{33}	z^{38}	z^{43}	z^{48}
1	1	z^{5}	z^{10}	z^{15}	z^{20}	z^{25}	z^{30}	z^{35}	z^{40}
$\times$	1	z^{5}	z^{10}	z^{15}	z^{20}	z^{25}	z^{30}	z^{35}	z^{40}

Challenge 1. Can you pinpoint where Sharp's solution relies on assumption 1?

You may also have noticed similarities between our solutions to the stamp problem and our discussion of Pick's formula for the area of lattice polygons in Chapter 2. This connection is made explicit in Problem 18.

6.4 Beyond Sylvester: The Stamp Theorem

Let us examine our pairings of numbers in arithmetic arrays more carefully.

Bijections

It will be helpful to formalize our notion of pairing first. A *bijection* between two sets is a list of pairs, where the first element is in the first set, and the second element is in the second set. Each element of the two sets is used once.

Example 2. (a) There is a natural bijection between the letters of the alphabet and the numbers from 1 to 26:

$$[\mathrm{A}, 1], [\mathrm{B}, 2], [\mathrm{C}, 3], \ldots, [\mathrm{Z}, 26].$$

(b) Our discovery of Sylvester's formula hinged on the bijection between the set of attainable numbers less than vw and the set of attainable number greater than vw that occur in the arithmetic array $A(v, w)$. The rule that governs the pairing is

$$n \longleftrightarrow 2vw - n.$$

The schematic

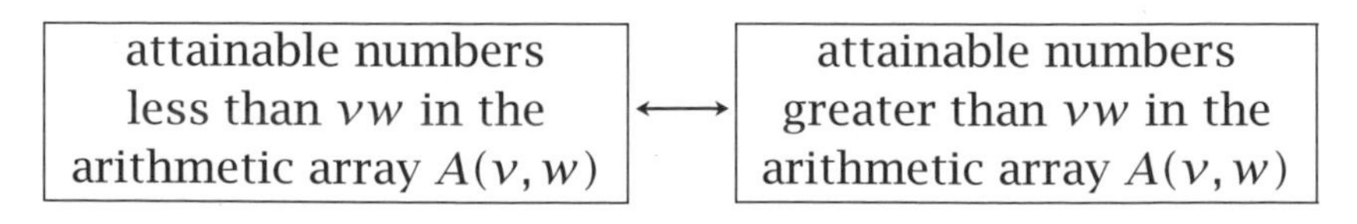

describes the two sets connected by the bijection.

Obviously, two sets must have the same number of elements for there to be a bijection between them. This fact allowed us to count the attainable numbers less than vw in the arithmetic array $A(v, w)$.

We now discuss a bijection that leads to a stronger version of Sylvester's formula—one that counts all of the unattainable numbers and identifies the largest one. For coprime denominations v and w, we define the parameters

$$U(v, w) = \text{the largest unattainable number}$$

$$\text{and} \quad N(v, w) = \text{the number of unattainable numbers.}$$

When the values of v and w are clear from context, we write U and N instead of $U(v, w)$ and $N(v, w)$. Our earlier work shows that

$$U(5, 8) = 27 \qquad \text{and} \qquad N(5, 8) = 14.$$

By the way, we are assuming that a largest unattainable number actually exists. Conceivably, some particular pair of denominations v and w leads to an infinite sequence of unattainable numbers. One of our tasks is to eliminate this possibility.

A Uniquely Attainable Pairing

Perhaps you noticed that the numbers listed in Table 6.1 are *uniquely attainable*. The equation $n = 5x + 8y$ has just one solution (x, y) for each attainable number n less than 40. There are two solutions for $n = 40$, namely, $(x, y) = (8, 0)$ and $(0, 5)$, but $n = 41, 42, 43$, and 44 are uniquely attainable. The 14 uniquely attainable numbers greater than 40 are

$$41,\ 42,\ 43,\ 44,\ 46,\ 47,\ 49,\ 51,\ 52,\ 54,\ 57,\ 59,\ 62,\ 67.$$

Table 6.4 gives the unique representations. It is easy to ver-

Table 6.4: Uniquely attainable numbers greater than 40

$41 = 5\times5 + 8\times2$	$51 = 5\times7 + 8\times2$
$42 = 5\times2 + 8\times4$	$52 = 5\times4 + 8\times4$
$43 = 5\times7 + 8\times1$	$54 = 5\times6 + 8\times3$
$44 = 5\times4 + 8\times3$	$57 = 5\times5 + 8\times4$
$46 = 5\times6 + 8\times2$	$59 = 5\times7 + 8\times3$
$47 = 5\times3 + 8\times4$	$62 = 5\times6 + 8\times4$
$49 = 5\times5 + 8\times3$	$67 = 5\times7 + 8\times4$

ify that none of the five consecutive numbers 68, 69, ..., 72 is uniquely attainable, and it follows that no larger number is uniquely attainable either. We listed the 14 numbers unattainable by 5 and 8 at the start of this chapter. All of them were less than 40:

$$1,\ 2,\ 3,\ 4,\ 6,\ 7,\ 9,\ 11,\ 12,\ 14,\ 17,\ 19,\ 22,\ 27.$$

The relationships between our two lists of 14 numbers leaps out at us. The positive number n is *un*attainable by 5 and 8 if and only if $n + 40$ is *uniquely* attainable. If you try other co-prime numbers v and w, you will always find a bijection given by the rule

$$n \longleftrightarrow n + vw$$

linking the two sets

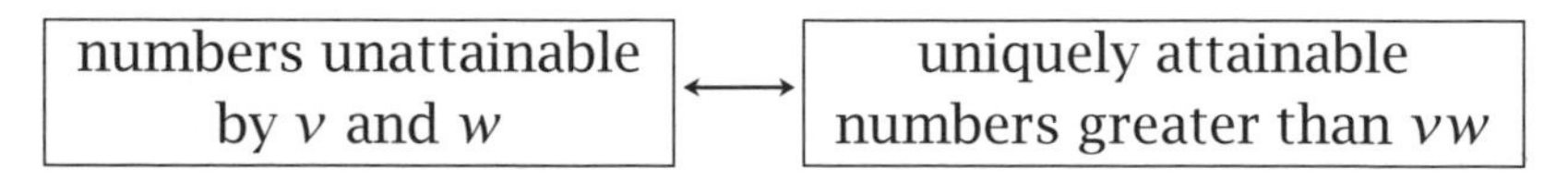

Diminished Arithmetic Arrays

The uniquely attainable numbers greater than vw are clustered together in the arithmetic array $A(v, w)$, as shown in Figure 6.3. It will be helpful to display the unattainable numbers inside a modified array.

40	45	50	55	60	65	70	75	80
32	37	42	47	52	57	62	67	72
24	29	34	39	44	40	54	59	64
16	21	26	31	36	41	46	51	56
8	13	18	23	28	33	38	43	48
0	5	10	15	20	25	30	35	40

uniquely attainable numbers greater than 40

Figure 6.3: The uniquely attainable numbers greater than 40 in the arithmetic array $A(5,8)$

The *diminished arithmetic array* $A^*(5,8)$ in Figure 6.4 is obtained from $A(5,8)$ by deleting the top row and rightmost column and subtracting 40 from each entry greater than 40. Now the shaded region includes the numbers unattainable by $v = 5$ and $w = 8$. The upper right corner contains the largest unattainable number $U = 27$. Note that each of the 40 numbers 0, 1, 2, ..., 39 occurs once in the diminished array.

32	37	2	7	12	17	22	27
24	29	34	39	4	9	14	19
16	21	26	31	36	1	6	11
8	13	18	23	28	33	38	3
0	5	10	15	20	25	30	35

$U = 27$

unattainable numbers

Figure 6.4: The diminished arithmetic array $A^*(5,8)$

The same idea works for arbitrary co-prime stamp denominations v and w. Form the *diminished arithmetic array* $A^*(v, w)$ by deleting the top row and rightmost column of the arithmetic array $A(v, w)$ and subtracting vw from every entry greater than vw. The diminished array has vw entries, and the largest unattainable number must appear in the upper right corner.

The Stamp Theorem

The diminished arithmetic array $A^*(v, w)$ is our key to improving Sylvester's result. The rest of our discussion depends on the following assumption.

Assumption 2. The numbers 0, 1, 2, ..., $vw - 1$ each occur once in the diminished arithmetic array $A^*(v, w)$ when v and w are co-prime.

The $(v - 1)(w - 1)/2$ attainable numbers less than vw are clustered in a region in the lower left corner of $A^*(v, w)$. The complementary region contains the unattainable numbers less than vw, and the cell $(x, y) = (w - 1, v - 1)$ in the extreme upper right corner clearly contains the largest of these, which is

$$\begin{aligned} U &= vx + wy \ - \ vw \\ &= v(w-1) + w(v-1) \ - \ vw \\ &= vw - v - w. \end{aligned}$$

This means that the numbers greater than $vw - v - w$ and less than vw are all attainable, giving us a run of at least v consecutive attainable numbers. It follows that all numbers larger than $vw - v - w$ are attainable. We have shown that Sylvester's restriction to unattainable numbers less than vw is unnecessary. Here is a summary of what we have learned.

Stamp theorem. Let the stamp denominations v and w be co-prime and greater than 1.

(a) The largest unattainable number is

$$U = vw - v - w.$$

(b) The number of unattainable numbers is

$$N = \frac{(v-1)(w-1)}{2} = \frac{U+1}{2}.$$

Part (a) is Sylvester's formula without the restriction to numbers less than vw.

The Arithmetic Array Theorem

The following result justifies the two assumptions we made en route to the stamp theorem.

Arithmetic array theorem.[2] Let v and w be co-prime numbers greater than 1.

(a) No two numbers in the arithmetic array $A(v, w)$ are equal, except that vw occurs in the upper left and the lower right corners.

(b) The numbers 0, 1, ..., $vw - 1$ each occur once in the diminished arithmetic array $A^*(v, w)$.

Assertion (a) justifies assumption 1, and (b) justifies assumption 2. To show that (a) is true, assume that the numbers in the distinct cells (x_1, y_1) and (x_2, y_2) are equal. Then

$$vx_1 + wy_1 = vx_2 + wy_2.$$

This relation can be manipulated to show that the distinct cells must be in the upper left and lower right corners. Problem 7 gives the details of the argument. The proof of assertion (b) is similar and is outlined in Problem 8.

[2]A theorem with the same name and similar proof applies to the arithmetic array $A(v, -w)$ in Chapter 5.

6.5 Chinese Remainders

We digress to connect our arithmetic arrays to a venerable result of number theory. We begin with a puzzle.

Question. When a teacher's class breaks into teams of five, there are four children left out, and when they break into teams of eight, two children are left out. How many children are in the class?

The diminished arithmetic array $A^*(5, 8)$ in Figure 6.5 provides a quick visual answer to the question. Each entry

32	37	2	7	12	17	22	27
24	29	34	39	4	9	14	19
16	21	26	31	36	1	6	11
8	13	18	23	28	33	38	3
0	5	10	15	20	25	30	35

Figure 6.5: The congruences $n \equiv 4 \pmod 5$ and $n \equiv 2 \pmod 8$

in the highlighted row is four more than a multiple of 5, and each entry in the highlighted column is two more than a multiple of 8. The number 34 in the intersection meets both conditions. So there are 34 children in the class. The number of children could also be $34 + 40 = 74$, or any number of the form $34 + 40k$ for $k = 0, 1, 2, \ldots$, but these larger numbers are less realistic for class sizes.

To answer the question for teams of five with $\hat{v}$ children left over and teams of eight with $\hat{w}$ children left over, we need only find the intersection of the appropriate row and column in the diminished arithmetic array $A^*(5, 8)$. Part (b) of our arithmetic array theorem guarantees that each number from 0 to 39 occurs in a unique row and column of $A^*(5,8)$, and thus there is a just one answer in that range for each pair of given remainders $\hat{v}$ and $\hat{w}$.

The same reasoning applied to the array $A^*(v,w)$ gives us a fundamental result in number theory, known in third-century China.

Chinese Remainder Theorem. Let v and w be co-prime, and let $\hat{v}$ and $\hat{w}$ be integers. Then there is a unique number n among $0, 1, \ldots, vw-1$ satisfying the two congruences

$$n \equiv \hat{v} \pmod{v} \qquad \text{and} \qquad n \equiv \hat{w} \pmod{w}.$$

We have used modular arithmetic to state the theorem. Recall that

$$n \equiv b \pmod{m}$$

means that $n-b$ is divisible by the *modulus* m. We say that n is *congruent* to b modulo m. When n and b are nonnegative, this implies that n and b leave the same remainder when divided by m.

Although the Chinese Remainder Theorem guarantees the existence of a number satisfying a pair of congruences, it does not tell us how to calculate the number. Of course, we could construct the diminished arithmetic array $A^*(v,w)$ and inspect the rows and columns, as we did above, but this process is cumbersome when v and w are large. A more efficient method relies on the co-primality theorem (discussed in Chapters 4 and 5). The following example illustrates the process.

Example 3. Let us find the smallest positive integer n satisfying the two congruences

$$n \equiv 38 \pmod{103} \qquad \text{and} \qquad n \equiv 17 \pmod{127}.$$

Example 2 of Chapter 4 shows that the co-primality relation

$$1 = 103x + 127y$$

is satisfied by $x = 37$ and $y = -30$. It follows that

$$\begin{aligned} 1 &\equiv 103 \times 37 \pmod{127} \\ \text{and} \qquad 1 &\equiv 127 \times (-30) \pmod{103}. \end{aligned}$$

Also,

$$\begin{aligned} 103 \times 37 \times 17 &\equiv 17 \pmod{127} \\ \text{and} \qquad 127 \times (-30) \times 38 &\equiv 38 \pmod{103}. \end{aligned}$$

It follows that the number

$$n = 103 \times 37 \times 17 \; + \; 127 \times (-30) \times 38 \; = \; -79993$$

satisfies our two congruences. In fact, for any integer k, the number

$$n = -79993 + (103 \times 127)k$$

satisfies the congruences, and we easily find that $n = 11574$ is the smallest positive integer of this form (corresponding to $k = 7$).

6.6 The Tabular Sieve

The *tabular sieve* generates all of the unattainable numbers for a given pair of co-prime denominations by sifting out the attainable numbers from the nonnegative integers.[3] The following observation forms the basis of the sieve.

Observation. If n is attainable by v and w, then so are $n + v$, $n + 2v$, $n + 3v$, ...

The sieve uses successive multiples of w for n.

[3]You may be familiar with the *sieve of Eratosthenes,* which operates in a similar manner to sift out the nonprimes and leave the primes.

Table 6.5: The tabular sieve for $v = 8$ and $w = 13$

0	8	16	24	32	40	48	56	64	72	80	88	96	···
1	9	17	25	33	41	49	57	65	72	81	89	97	···
2	10	18	26	34	42	50	58	66	73	82	90	98	···
3	11	19	27	35	43	51	59	67	74	83	91	99	···
4	12	20	28	36	44	52	60	68	75	84	92	100	···
5	13	21	29	37	45	53	61	69	76	85	93	101	···
6	14	22	30	38	46	54	62	70	78	86	94	102	···
7	15	23	31	39	47	55	63	71	79	87	95	103	···

Example 4. Here is how the sieve works when $v = 8$ and $w = 13$. Place the nonnegative integers in a table with 8 rows, as shown in Table 6.5. The sieve will shade the attainable numbers. We work our way through the table's entries 0, 1, 2, ... one at a time. When an unshaded multiple of 13 is encountered, we shade it, as well as all larger entries in the same row. The first multiple of 13 we encounter is 0, and so we shade 0 and the other numbers (8, 16, 24, ...) in the top row. The next multiple of 13 is 13 itself, and we shade 13, 21, 29, ... Then we shade 26, 34, 42, ... We stop when no larger unshaded numbers remain. The sieve produces the shaded and unshaded regions in Table 6.5. It is clear that the largest unattainable number is $U(8, 13) = 83$, and a direct count of the unshaded entries shows that the number of unattainable numbers is $N(8, 13) = 42$. These values agree with the stamp theorem.

In general, the tabular sieve for v and w starts with a table of v rows, say, and steps through the numbers 0, 1, 2, ... Each multiple of w is shaded as it is encountered, as are the larger numbers in the same row. The unattainable numbers remain unshaded. The largest unattainable number U

is readily located within the final unshaded configuration.

Perhaps you noticed that the fingers of the shaded and unshaded regions in the tabular sieve interlock symmetrically. Problem 15 explains the details of this symmetry and the underlying bijection.

6.7 McNuggets and Coin Exchanges

There is a popular version of a natural extension of the stamp problem to more than two denominations.

McNuggets problem. If McDonald's sells Chicken McNuggets[4] in boxes of 6, 9, and 20, what is the largest exact number of McNuggets that cannot be ordered?

Algebraically, the problem asks for the largest positive number n such that

$$n = 6x + 9y + 20z$$

has no solution. The variables x, y, and z are assumed to be nonnegative integers. The answer is $n = 43$, which can be verified by considering various cases. A more methodical approach will be given later.

It is customary to state the general problem in a numismatic setting rather than a gustatory or philatelic one.

The coin exchange problem (Frobenius). Let $v_1, v_2, \ldots, v_m$ be m co-prime numbers. Find the largest number n for which the equation

$$n = v_1x_1 + v_2x_2 + \cdots + v_mx_m$$

has no solution in nonnegative integers $x_1, x_2, \ldots, x_m$.

[4]McNuggets is a registered trademark of McDonald's Restaurant.

If our coins have denominations v_1, v_2, $\ldots$, v_m, then the answer to the coin exchange problem is the largest amount of money for which we cannot make change. Part (a) of the stamp theorem solves the case with $m = 2$ denominations.

The denominations must satisfy

$$\gcd(v_1, v_2, \ldots, v_m) = 1$$

for the coin exchange problem to make sense, and we always assume this co-primality condition holds. We also assume that the denominations are in increasing order and that the smallest denomination is not 1:

$$1 < v_1 < v_2 < \cdots < v_m.$$

The coin exchange problem bears the name of Ferdinand Georg Frobenius (1849–1917), a versatile German mathematician who popularized the problem through his lectures. In his honor, the largest unattainable number with coins of denominations v_1, v_2, $\ldots$, v_m is called the *Frobenius number.* We denote the Frobenius number by

$$U(v_1, v_2, \ldots, v_m),$$

or simply U when the coin denominations are clear from context. The coin exchange problem asks for the value of $U(v_1, v_2, \ldots, v_m)$.

Example 5. (a) For the case $m = 2$, we have seen that

$$U(v_1, v_2) = v_1 v_2 - v_1 - v_2$$

However, no one has found a simple formula for the Frobenius number for more than two denominations.

(b) It is not difficult to verify that $U(3, 5, 7) = 4$. The only three unattainable numbers are 1, 2, and 4.

(c) The answer to the McNuggets problem is $U(6, 9, 20) = 43$.

The Tabular Sieve Again

In the absence of a simple formula, it is natural to seek algorithms that compute the Frobenius number for any given list of denominations $v_1, v_2, \ldots, v_m$. One such algorithm is an extension of the sieve we used for two denominations.

Table 6.6: The tabular sieve and the coin exchange problem

0	6	12	18	24	30	36	42	48	⋯
1	7	13	19	25	31	37	43	49	⋯
2	8	14	20	26	32	38	44	50	⋯
3	9	15	21	27	33	39	45	51	⋯
4	10	16	22	28	34	40	46	52	⋯
5	11	17	23	29	35	41	47	53	⋯

Example 6. Table 6.6 illustrates the tabular sieve for $v_1 = 6$, $v_2 = 9$, $v_3 = 20$ (the McNuggets problem). We place the nonnegative integers in a table with $v_1 = 6$ rows and walk through the numbers 0, 1, 2, ..., shading any number attainable by $v_2 = 9$ and $v_3 = 20$, as well as the larger numbers in the same row. So first we shade all numbers 0, 6, 12, ... in the top row. Next we shade 9, 15, 21, ... and then 20, 26, 32, ... Eventually all six rows are accounted for, and we confirm that $U(6, 9, 20) = 43$.

A Coin Theorem

Neither part of the stamp theorem generalizes neatly to more than two denominations. However, the two formulas in the stamp theorem do extend to analogous *inequalities.*

Coin theorem. Let the coin denominations satisfy $1 < v_1 < v_2 < \cdots < v_m$ and $\gcd(v_1, v_2, \ldots, v_m) = 1$.

(a) The largest amount for which we cannot make change satisfies

$$U \le (\nu_1 - 1)(\nu_m - 1) - 1.$$

(b) The number of unattainable amounts satisfies

$$N \ge \frac{U+1}{2}.$$

The upper bound for the Frobenius number in part (a) follows from the stamp theorem if ν_1 and ν_m are co-prime but is more difficult to prove if ν_1 and ν_m have a common divisor greater than 1. The bound was established by Issai Schur (1875–1941), who was a student of Frobenius and a prominent mathematician in his own right.

Another Pairing

There is a clever way to see why part (b) of the coin theorem is true even though we do not have explicit formulas for N and U. Each of the $U + 1$ pairs

$$[0, U],\ [1, U-1],\ [2, U-2], \ldots, [U, 0]$$

must contain at least one unattainable number. For if both numbers in the pair $[n, U - n]$ were attainable, then their sum $n + (U - n) = U$ would also be attainable, contrary to the definition of U as the largest unattainable number. It follows that $N \ge (U + 1)/2$. Equality holds exactly when each pair has one attainable number and one unattainable number. This always happens with $m = 2$ denominations, in which case we have the bijection

$$n \longleftrightarrow U - n$$

for the sets

$$\boxed{\begin{array}{c}\text{attainable numbers}\\ \text{among } 0, 1, 2, \ldots, U\end{array}} \longleftrightarrow \boxed{\begin{array}{c}\text{unattainable numbers}\\ \text{among } 0, 1, 2, \ldots, U\end{array}}.$$

This bijection sheds more light on the equality

$$N = \frac{U+1}{2}$$

from part (b) of the stamp theorem.

6.8 Sylver Coinage

John Horton Conway, one of the world's great contemporary mathematicians, invented a game inspired by the coin exchange problem. He called the game *sylver coinage* in honor of Sylvester. Two players, Alice and Bob, take turns naming positive integers out loud. Alice goes first and selects any positive integer to start the game. Thereafter, a player must name a number that is unattainable by the numbers already selected. The person forced to name the number 1 loses.

What strategies should the players adopt? Who wins if Alice and Bob both play optimally? The answers are not obvious. It is helpful to look at some small examples to get a feel for the game.

Example 7. (a) Suppose Alice leads off with 3. Then Bob immediately replies with 2. Alice must select 1, the only remaining unattainable number, and Bob wins. Similarly, if Alice leads off with 2, then Bob selects 3 and wins. So Alice should not lead off with 2 or 3.

(b) Suppose Alice leads off with 4, and Bob replies with 5. Alice can then win by selecting 11. For then Bob must select a number from the list 1, 2, 3, 6, 7. If he selects 7, then she selects 6; if he selects 6, she selects 7. It follows that Bob will select 2 or 3 before Alice does, and she wins by following the strategy in (a). So Bob should not select 5 if Alice leads off with 4. Nor should he select 4 if Alice leads off with 5.

Sylver coinage is vastly more complicated than Example 7 suggests. For instance, suppose Alice leads off with 8, and Bob replies with 30, and Alice selects 34. It has been shown that Bob can guarantee a win, but only if he selects the five-digit number 49337 next. Any other response by Bob loses if Alice plays optimally. Such facts are discovered through a mixture of theory and computer explorations. In general, it may be extremely difficult to determine the best number to select at a particular stage of sylver coinage.

Alice Steals Bob's Strategy

A *strategy-stealing* argument shows that Alice wins sylver coinage if she plays optimally from the start. The following example illustrates the basic idea with numbers from the McNuggets problem.

Example 8. Suppose Alice leads off with 6, Bob replies with 9, and then Alice selects 20. Bob will then select one of the $N = 22$ unattainable (unshaded) numbers in Table 6.6. Since the largest unattainable number is $U = 43 = 2N + 1$, each of the 22 mated pairs

$$[0,43],\ [1,42],\ [2,41],\ \ldots,\ [21,22]$$

contains one attainable number and one unattainable number. If selecting 43 is a winning strategy for Bob, then he should select 43, of course. Suppose, on the other hand, that 43 is a losing move for Bob. This means Alice can win by replying to 43 with some number $\overline{n}$ less than 43. With this observation in mind, Bob realizes he can steal Alice's strategy by immediately selecting $\overline{n}$ himself. This selection excludes 43 from play (since $\overline{n}$ and its attainable mate sum to 43) and places Alice in the same predicament confronted by Bob had he selected 43 and Alice chosen $\overline{n}$.

We have shown that if 6, 9, and 20 are the first three numbers selected, then Bob can force a win. But we have not determined whether Bob should select 43 or some smaller number on his next move.

Alice can put herself in a position to steal Bob's strategy from the outset by leading off with 5. Suppose that Bob replies with b. Then 5 and b must be co-prime, and the largest number available to Alice is $4b - 5$ by the stamp theorem. Each of the pairs

$$[0, 4b - 5], [1, 4b - 6], [2, 4b - 7], \ldots, [2b - 3, 2b - 2]$$

has one attainable number and one unattainable number. If Alice cannot win by selecting $4b - 5$, then she can win by stealing Bob's response to $4b - 5$.

We have established the most important result about sylver coinage.

Sylver theorem. Alice wins sylver coinage if she plays optimally. For instance, she can lead off with 5.

The elegant strategy-stealing argument suffers from a practical drawback. It does not give Alice a specific winning retort to each of Bob's replies. For instance, suppose Alice leads off with 5. Then Bob replies with 666^{999} and smiles devilishly as he waits for Alice to figure out an optimal response.

6.9 Notes and References

See the award-winning biography by Parshall [8] for a detailed account of Sylvester's life and work. Sylvester's first discussion of the stamp problem appears on page 134 of an article [13] published in 1882. Sharp's solution to Sylvester's problem is reproduced on page 104 of [9]. Sharp's polynomials are an instance of the use of generating functions in discrete mathematics; see the texts [4] and [10], e.g., for introductions to this vast topic.

Topics related to Sylvester's stamp formula are also covered Chapter 13 of Honsberger's book [5]. Chapter 15 of Schumer's book [11] includes an accessible treatment of some of Sylvester's contributions to the partition problem.

The article by Owens [7] contains an excellent discussion of the coin exchange problem and includes many references not listed here. Hundreds of references occur in the monograph devoted to the coin exchange problem by Ramírez Alfonsín [9]. The inequality in part (a) of the coin theorem is attributed to Schur in [3], while (b) is by Nijenhuis and Wilf [6]. The circle of lights algorithm described in Problem 16 was devised by Wilf [14]. The recent electronic article [1] presents impressive new algorithms to compute Frobenius numbers.

Sylver coinage and strategy-stealing arguments are discussed in the classic book [2] and in the more recent article [12]. Sicherman's site

`www.monmouth.com/~colonel/sylver/`

contains a wealth of information about sylver coinage.

1. Beihoffer, D., Hendry, J., Nijenhuis, A., and Wagon, S., Faster algorithms for Frobenius numbers, *Electronic Journal of Combinatorics* **12** (2005), #R27 38 pp (electronic).
2. Berlekamp, Elwyn R., Conway, John H., and Guy, Richard K., *Winning Ways for Your Mathematical Plays.* Academic Press, London, 1985.
3. Brauer, A., On a problem of partitions, *American Journal of Mathematics* **64** (1942), 299–312.
4. Grimaldi, Ralph P., *Discrete and Combinatorial Mathematics: An Applied Introduction,* 5th ed. Addison Wesley Longman, Reading, Massachusetts, 2004.
5. Honsberger, Ross, *Mathematical Gems II.* Mathematical Association of America, Washington, DC, 1976.
6. Nijenhuis, A., and Wilf, H. S., Representation of integers by linear forms in non-negative integers, *Journal of Number Theory* **4** (1972), 98–106.
7. Owens, R. W., An algorithm to solve the Frobenius problem, *Mathematics Magazine* **76** (2003), 264–275.
8. Parshall, Karen Hunger, *James Joseph Sylvester: Jewish Mathematician in a Victorian World.* Johns Hopkins University Press, Baltimore, 2006.
9. Ramírez Alfonsín, J. L., *The Diophantine Frobenius Problem.* Oxford University Press, New York, 2005.
10. Roberts, Fred S., and Tesman, Barry, *Applied Combinatorics,* 2nd ed. Prentice Hall, Upper Saddle River, New Jersey, 2003.
11. Schumer, Peter D., *Mathematical Journeys.* Wiley Interscience, Hoboken, New Jersey, 2004.

12. Sicherman, G., Theory and practice of Sylver Coinage, *Integers: Electronic Journal of Combinatorial Number Theory* **2** (2002), #G02 11 pp (electronic).
13. Sylvester, J. J., On subinvariants, i.e. semi-invariants to binary quantics of an unlimited order, *American Journal of Mathematics* **5** (1882), 119–136.
14. Wiilf, H. S., A circle-of-lights algorithm for the "money-changing problem," *American Mathematical Monthly* **85** (1978), 562–565.

6.10 Problems

1. Exactly 63 numbers are unattainable by v and w.

   (a) What are the possible values of v and w?

   (b) What are v and w if 24 is unattainable?

2. The largest number unattainable by v and w is 63. Must 30 be attainable?

3. Interpret the formulas in parts (a) and (b) of the stamp theorem for the forbidden case $v = 1$.

4. Find the smallest positive integer that leaves remainders 2, 3, and 4 when divided by 5, 7, and 9, respectively.

5. Find the Frobenius numbers.

   (a) $U(6, 10, 15)$

   (b) $U(10, 12, 15)$

   (c) $U(15, 16, 17)$

   (d) $U(9, 10, 11, 12)$

6. Let v and w be co-prime numbers greater than 1. Find a formula for the smallest attainable number S that does *not* occur in the arithmetic array $A(v, w)$. For example, Figure 6.1 tells us that $S = 53$ when $v = 5$ and $w = 8$.

7. This problem gives the details of a proof of part (a) of the arithmetic array theorem. Let v and w be co-prime numbers greater than 1. Assume that the numbers in the distinct cells (x_1, y_1) and (x_2, y_2) of the arithmetic array $A(v, w)$ are equal. Then

$$vx_1 + wy_1 = vx_2 + wy_2.$$

   (a) Explain why $x_1 \neq x_2$. Henceforth suppose that $x_1 < x_2$.

   (b) Show that
   $$\frac{y_1 - y_2}{x_2 - x_1} = \frac{v}{w}.$$

   (c) Explain why the fraction v/w is in lowest terms.

   (d) Explain why $x_2 - x_1$ is a positive multiple of w, and $y_1 - y_2$ is a positive multiple of v.

   (e) Deduce that $x_2 = w$ and $x_1 = 0$. Hint: x_1 and x_2 are in the list $0, 1, \ldots, w$.

   (f) Deduce that $y_1 = v$ and $y_2 = 0$.

   (g) Deduce that the only two equal numbers in the arithmetic array $A(v, w)$ occur in the upper left and lower right corners.

8. This problem outlines a proof of part (b) of the arithmetic array theorem.

   (a) Explain why it is sufficient to show that no two numbers in the diminished arithmetic array $A^*(v, w)$ are equal.

   (b) Assume that the same number appears in cells (x_1, y_1) and (x_2, y_2). Explain why exactly one of the numbers in these cells must have been diminished by vw in passing from $A(v, w)$ to $A^*(v, w)$. Hint: Use part (a) of the arithmetic array theorem.

   (c) Analyze the equation

$$vx_1 + wy_1 = vx_2 + wy_2 - vw$$

to complete the argument. Remember that x_1 and x_2 are in the list $0, 1, \ldots, w-1$, while y_1 and y_2 are in the list $0, 1, \ldots, v-1$.

9. Here is an excerpt from Sylvester's article of 1882.

> The number of integers less than pq and containing neither p nor q is $(p-1)(q-1)$, and if every two of these which are supplementary to one another (I mean whose sum is pq) be made into a pair, it is an easily demonstrable, but by no means an unimportant fact, that one of the pair will be a compound and the other a non-compound of p and q. Hence the total number of non-compounds is $(p-1)(q-1)/2$, and Therefore, the total number of solutions of $px + qy < pq$ will be the remainder when the above is subtracted from pq.

   (a) What does "containing neither p nor q" mean?

   (b) What is a "compound"?

   (c) Explain why the "easily demonstrable" fact is true.

   (d) Explain why the smallest compound containing neither p nor q is $p+q$ and deduce that the largest noncompound less than pq is $pq-p-q$. This is Sylvester's formula.

10. Let v and w be co-prime numbers with $1 < v < w$. Find a formula for the second largest unattainable number.

11. (a) Suppose Alice leads off a game of sylver coinage with 6, Bob replies 5, and Alice names 4. What is Bob's best response?

   (b) Alice leads off a game of sylver coinage with 35. Give two winning replies for Bob.

   (c) Alice leads off a game of sylver coinage with 10. Give a winning reply for Bob.

12. (a) Show that Alice wins sylver coinage if she leads off with any prime greater than 3 and plays optimally.

    (b) Explain how Bob wins if Alice leads off with any number possessing at least two prime divisors greater than 3.

13. Let d be the greatest common divisor of v and w.

    (a) Explain why infinitely many numbers are unattainable if $d > 1$.

    (b) Show that the number of unattainable numbers less than vw and divisible by d is

$$\frac{(v-d)(w-d)}{2d^2}.$$

    (c) Show that the number of unattainable numbers less than vw is

$$\frac{(v-d)(w-d)}{2d^2} + \left(\frac{d-1}{d}\right) vw.$$

14. Let v and w be co-prime numbers greater than 1.

    (a) Show that sum of the numbers in the diminished arithmetic array $A^*(v,w)$ is

$$\frac{vw(vw-1)}{2}.$$

    (b) Show that the sum of the numbers in the arithmetic array $A(v,w)$ is

$$vw(v+1)(w+1).$$

    Hint: Exploit symmetry to find the average value in the array $A(v,w)$.

15. This problem investigates the symmetry of the interlocking shaded and unshaded fingers that occur when the tabular sieve is carried out for two co-prime numbers.

Table 6.7: Symmetry and the tabular sieve for $v = 8$ and $w = 13$

1	9	17	25	33	41	49	57	65	73	81	89
2	10	18	26	34	42	50	58	66	74	82	90
3	11	19	27	35	43	51	59	67	75	83	91
4	12	20	28	36	44	52	60	68	76	84	92
5	13	21	29	37	45	53	61	69	77	85	93
6	14	22	30	38	46	54	62	70	78	86	94
7	15	23	31	39	47	55	63	71	79	87	95

(a) Table 6.7 focuses on a certain sub-rectangle within the tabular sieve for $v = 8$ and $w = 13$. (Compare with Table 6.5). A 180° spin about the center of the rectangle interchanges the shaded and unshaded regions and gives us a bijection. Complete the rule for the bijection

$$n \longleftrightarrow ?$$

and fill in the box with a suitable set:

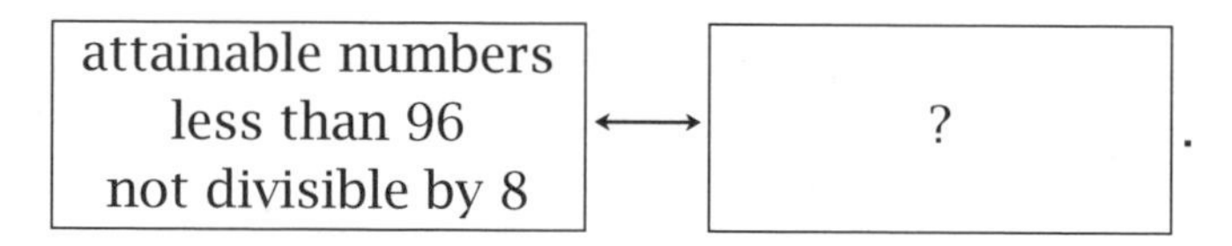

(b) Let v and w be co-prime numbers with $v < w$. Generalize part (a) to the tabular sieve with v rows.

(c) Use part (b) to confirm that the the number of unattainable numbers for v and w is $N = (v-1)(w-1)/2$.

16. The *circle of lights algorithm* finds the Frobenius number $U(v_1, v_2, \ldots, v_m)$. Start with a circle of v_m bulbs, all of which are off. Switch on one bulb, label it $n = 0$, and let $u = 0$. Proceed clockwise around the circle, increasing n by 1 at each bulb. If the bulb is off, then switch it on if condition A is met; let $u = n$ if the condition is not met. If the bulb is

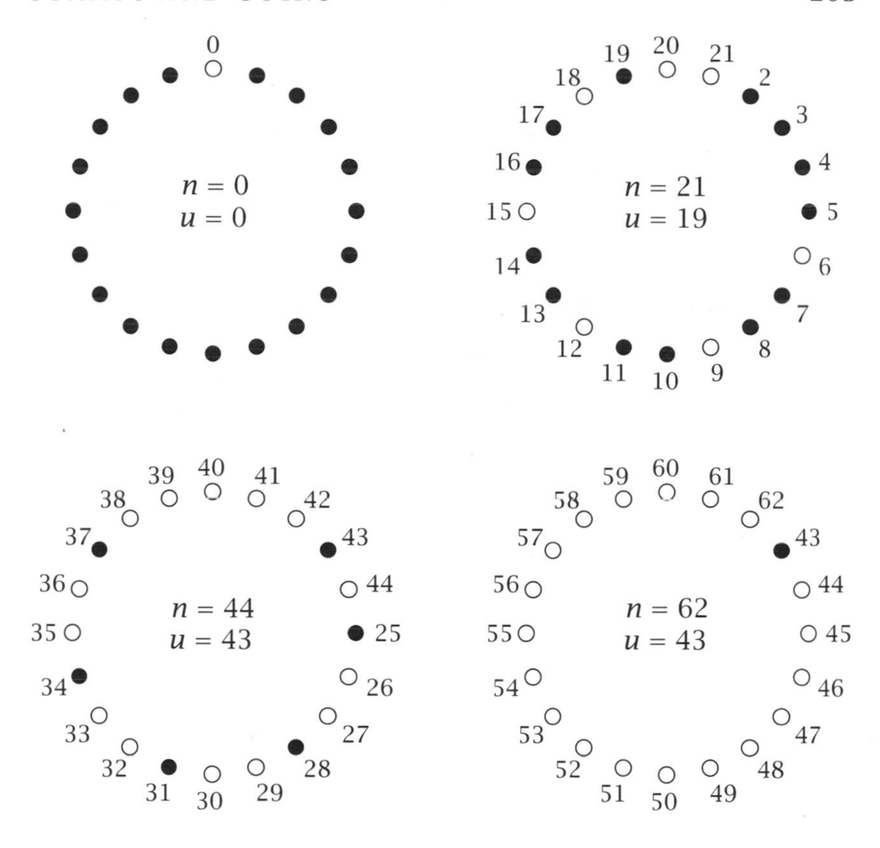

Figure 6.6: The circle of lights algorithm

on, then simply move to the next bulb. The algorithm stops when condition B is met. Figure 6.6 depicts the algorithm's computation of $U(6, 9, 20)$.

(a) What is condition A?

(b) What is condition B?

(c) Carry out the algorithm to compute $U(10, 12, 15)$.

(d) Explain why $u = U(v_1, v_2, \ldots, v_m)$ when the algorithm stops.

17. This problem explains why every game of sylver coinage must end eventually. In other words, it is impossible for Alice and Bob to conspire to choose an infinite list of numbers. The key is to consider the list of numbers $g_1, g_2, \ldots$, where g_k is the greatest common divisor of the first k numbers named.

    (a) Explain why $g_1 \geq g_2 \geq g_3 \geq \cdots$.

    (b) Explain why $g_k = g_{k+1} = g_{k+2} = \cdots$ for some k.

    (c) Show that if $g_k = 1$ for some k, then the game must end eventually. Hint: Use the coin theorem.

    (d) Show that it is impossible to have $g_k > 1$ for all k. Hint: Division reduces the situation to part (c).

18. This problem explains how to deduce Sylvester's formula from Pick's formula (Chapter 2) for the area of a lattice polygon with L lattice points and B boundary lattice points:

$$\text{area of lattice polygon} = L - \tfrac{1}{2}B - 1.$$

Let v and w be co-prime numbers greater than 1. Our strategy is to determine the total number of lattice points L in the triangle with vertices $(0,0)$, $(0,v)$, and $(w,0)$. These lattice points are closely related to the numbers attainable by v and w. Figure 6.7 illustrates the case with $v = 5$ and $w = 8$.

    (a) Show that the number of boundary lattice points of the triangle is $B = v + w + 1$.

    (b) Use Pick's formula to deduce that the total number of lattice points of the triangle is

$$L = \frac{(v+1)(w+1)}{2} + 1.$$

    (c) Explain why each of the L lattice points except for the two vertices $(0,v)$ and $(w,0)$ corresponds to an attainable number for v and w less than vw. Hint: What region

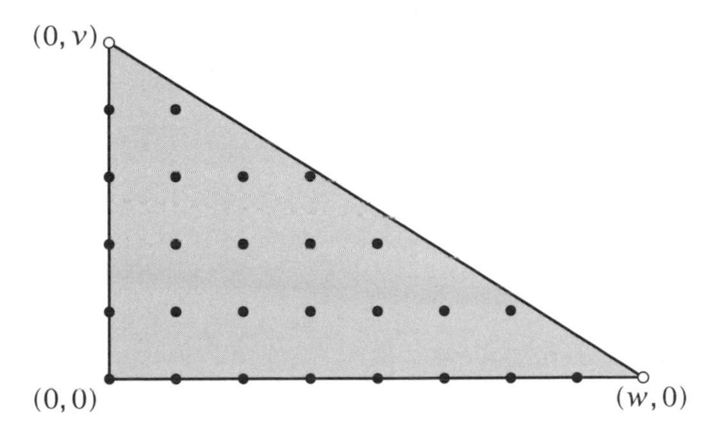

Figure 6.7: From Pick to Sylvester

is defined by the the inequalities

$$vx + wy < vw, \qquad x \geq 0, \qquad y \geq 0?$$

(d) Deduce Sylvester's formula for the number of unattainable numbers less than vw:

$$N = \frac{(v-1)(w-1)}{2}.$$

(e) On what assumption does the above method rely?

Mathematics is not a book confined within a cover and bound between brazen clasps, whose contents it needs only patience to ransack . . . its possibilities are as infinite as the worlds which are forever crowding in and multiplying upon the astronomer's gaze.

J. J. SYLVESTER

7

Primes and Squares: Quadratic Residues

I think the prime reason for existence,
for living in this world, is discovery.
JAMES DEAN

7.1 Primes and Squares

In the eighteenth century, mathematicians discovered several striking relationships between the primes[1]

$$2, 3, 5, 7, 11, 13, 17, 19, 23, 29, 31, 37, 41, 43, \ldots$$

and the squares

$$0, 1, 4, 9, 16, 25, 36, 49, 64, 81, 100, 121, 144, \ldots.$$

Their discoveries arose from an innocuous question.

Question. What nonzero remainders can occur when we divide each square by a given prime?

When we divide a nonnegative integer n by a positive integer p, we get one of the remainders $0, 1, 2, \ldots, p-1$. However, if n is a square and p is a prime, only some of the remainders actually occur. For instance, when we divide

[1] A prime is a number with exactly two factors, namely, itself and 1.

the squares 0, 1, 4, 9, ... by the prime 11, the successive remainders are

$$\underbrace{0, 1, 4, 9, 5, 3, 3, 5, 9, 4, 1,}_{\text{block of eleven}}\underbrace{0, 1, 4, 9, 5, 3, 3, 5, 9, 4, 1,}_{\text{block of eleven}} \ldots.$$

The blocks of eleven repeat, and the only nonzero remainders are 1, 3, 4, 5, and 9. These five numbers are the *quadratic residues* modulo 11.

Our goal in this chapter is to discover and explain properties of quadratic residues. Our discussion culminates in the beautiful Law of Quadratic Reciprocity. We will also find a surprising link between quadratic residues and the computer line drawing problem discussed in Chapter 4.

The study of quadratic residues was rooted in curiosity about the fundamental properties of the integers. No specific application was contemplated or even deemed possible. For over two millenia, number theory stood as the quintessential branch of pure mathematics, isolated from the traditional real-world applications of mathematics in the physical sciences. The development of sophisticated communication systems in the twentieth century gave rise to new types of mathematical problems dealing with the transmission, sharing, and withholding of information. Happily, the number theory developed earlier for curiosity's sake was ideally suited to solve many of the new problems. Our adventures in this chapter include glimpses of the use of quadratic residues in coding theory and in cryptography.

7.2 Quadratic Residues Are Squares

We introduced quadratic residues as the possible nonzero remainders when the squares are divided by a given prime. Modular arithmetic lets us bypass the division and work directly with the remainders in many situations. Recall that a

is *congruent* to b modulo m, written

$$a \equiv b \pmod{m},$$

provided $a - b$ is divisible by the positive integer m. When a and b are nonnegative integers, then $a \equiv b \pmod{m}$ if and only if a and b leave the same remainder when divided by m. The *modulus* m will be a prime in most of our discussions. Congruences can be manipulated much like ordinary equalities, as we did in Chapter 5.

A quadratic residue is a nonzero square modulo a given prime. More formally, the integer v is a *quadratic residue* modulo the prime p provided v is not divisible by p, and some integer x satisfies the quadratic congruence

$$x^2 \equiv v \pmod{p}.$$

If the congruence has no solution, then v is a *quadratic nonresidue* modulo p. The number 0 is neither a quadratic residue nor a nonresidue. If $v \equiv v' \pmod{p}$, then v is a quadratic residue if and only if v' is, and so we often assume that $1 \leq v \leq p - 1$ when dealing with quadratic residues.

Example 1. (a) The number 1 is a quadratic residue modulo every prime since the congruence $x^2 \equiv 1 \pmod{p}$ clearly has the solutions $x = \pm 1$.

(b) Quadratic residues are not very interesting modulo $p = 2$. The only quadratic residue is 1. There are no quadratic nonresidues. For this reason, we usually restrict our attention to quadratic residues modulo odd primes.

(c) The five quadratic residues modulo 11 are 1, 3, 4, 5, and 9.

(d) Table 7.1 shows that the six quadratic residues modulo 13 are 1, 3, 4, 9, 10, and 12. Note that the last six quadratic residues in the table duplicate the first six in reverse order, giving us a palindromic sequence.

Table 7.1: Quadratic residues modulo 13

x	1	2	3	4	5	6	7	8	9	10	11	12
x^2	1	4	9	3	12	10	10	12	3	9	4	1

To find the quadratic residues modulo the odd prime p, we can compute the $(p-1)/2$ squares

$$1^2, 2^2, \ldots, \left(\frac{p-1}{2}\right)^2$$

modulo p. The congruence

$$(p-x)^2 \equiv p^2 - 2px + x^2 \equiv x^2 \pmod{p}$$

assures us that the squares of $(p+1)/2, \ldots, p-2, p-1$ duplicate the squares of $1, 2, \ldots, (p-1)/2$ in reverse order modulo p.

The Two Quadratic Residue Questions

There are two fundamental questions about quadratic residues.

1. Is the integer v a quadratic residue modulo a given prime p?
2. For which primes p is the given integer v a quadratic residue modulo p?

Less formally, the first question asks whether the square of an integer can be v more than a multiple of p. A straightforward solution is available. Simply check whether v is congruent to one of the squares $1^2, 2^2, \ldots, ((p-1)/2)^2$ modulo p. This direct computational method was used to construct Table 7.2, which lists the quadratic residues modulo the odd primes less than 60. The process becomes more tedious as the primes get larger. The second question does not succumb to a direct computational assault because there are infinitely many primes to check for each $\nu > 1$. We will see better ways to attack both questions later.

Some suggestive patterns can be discerned within the seemingly random data in Table 7.2. For instance, it happens that $p - 1$ is a quadratic residue modulo each prime

$$p = 5, 13, 17, 29, 37, 41, 53, \ldots$$

Table 7.2: Quadratic residues modulo odd primes less than 60

3	**5**	**7**	**11**	**13**	**17**	**19**	**23**	**29**	**31**	**37**	**41**	**43**	**47**	**53**	**59**
1	1	1	1	1	1	1	1	1	1	1	1	1	1	1	1
	4	2	3	3	2	4	2	4	2	3	2	4	2	4	3
		4	4	4	4	5	3	5	4	4	4	6	3	6	4
			5	9	8	6	4	6	5	7	5	9	4	7	5
			9	10	9	7	6	7	7	9	8	10	6	9	7
				12	13	9	8	9	8	10	9	11	7	10	9
					15	11	9	13	9	11	10	13	8	11	12
					16	16	12	16	10	12	16	14	9	13	15
						17	13	20	14	16	18	15	12	14	16
							16	22	16	21	20	16	14	16	17
							18	23	18	25	21	17	16	17	19
								24	19	26	23	21	17	24	20
								25	20	27	25	23	18	25	21
								28	25	28	31	24	21	28	22
									28	30	32	25	24	29	25
										33	33	31	25	36	26
										34	36	35	27	37	27
										36	37	36	28	38	28
											39	38	32	40	29
											40	40	34	42	34
												41	36	43	36
													37	44	41
													42	46	44
														47	46
														49	48
														52	49
															51
															53
															57

that is one more than a multiple of 4. Also, 2 appears to be a quadratic residue modulo the primes

$$p = 7, 17, 23, 31, 41, 47, \ldots$$

that are one more or one less than a multiple of 8. These two observations hold in general and are known as the *supplementary laws* for quadratic reciprocity.

Supplementary laws. Let p be an odd prime.

1. The number -1 is a quadratic residue modulo p if and only if $p \equiv 1 \pmod{4}$.
2. The number 2 is a quadratic residue modulo p if and only if $p \equiv \pm 1 \pmod{8}$.

We replaced $p-1$ with -1 in the first supplementary law. This is allowed because $p - 1 \equiv -1 \pmod{p}$.

7.3 Errors: Detection amd Correctipn

You are able to read the title of this section because the English language provides a context in which the two typographical errors are easy to correct.[2] Modern electronic media and communication systems lack a similar context because the "words" are seemingly arbitrary strings of symbols. For instance, suppose the words in Table 7.3, composed of the two symbols "+" and "−", represent the 16 possible shades of gray for each pixel in an image captured by a space probe. The entire image is transmitted to Earth by a long sequence of such words. Unfortunately, errors are sometimes introduced during transmission (say, by electromagnetic anomalies in the communication equipment or in Earth's atmosphere), switching a "+" to a "−" or vice versa.

[2]In fact, you may not have even noticed the errors because the human brain often makes the corrections automatically.

A single error in any word changes the shade for a pixel, diminishing the quality of the received image. We are faced with a problem in *coding theory*, the study of methods to transmit information reliably over imperfect communication channels.

Table 7.3: Sixteen words

$++++$	$+-++$	$-+++$	$--++$
$+++-$	$+-+-$	$-++-$	$--+-$
$++-+$	$+--+$	$-+-+$	$---+$
$++--$	$+---$	$-+--$	$----$

An *error-correcting code* creates a context for each word, which helps us detect and correct the errors and accurately reconstruct the image captured by the probe. Error-correcting codes are used throughout the computer industry (to store data and music on compact disks, for instance) as well as in telecommunication.

Our choice of the two symbols "+" and "−" agrees with our upcoming notation for quadratic residues, rather than the symbols "0" and "1" traditionally used in computer applications. All that matters is that our words use two distinguishable symbols.

Repetition Codes

In a *repetition code,* we transmit each 4-symbol word three times in succession as a 12-symbol *code word.* The *majority rules* scheme lets us correct a single error in any code word.

Example 2. Suppose we receive the 12 symbols

$$++-+++-+++--.$$

The third block of four symbols disagrees with the first two blocks in the last position. So we switch the symbol in that position from "−" and "+" to deduce the transmitted code word.

If two errors occur in the same position in two different blocks of our repetition code, then we are unable to reconstruct the transmitted code word correctly.

Parameters of a Code

The most important parameters of a code are its length, its size, and the number of errors it can correct in each code word. The *length* is the number of symbols in each code word, and the *size* is the number of code words. The repetition code described previously has length 12, size 16, and corrects one error.

For faster communication we prefer the length of the code to be small. Shorter code words also require less energy to transmit, an important consideration in space communication. But to distinguish among many shades of gray the code's size should be large. Of course, we want to correct as many errors as we can in each code word. These preferences are mutually incompatible. If we want to have more code words or correct more errors, then longer code words are required. The longer code words take longer to transmit but produce higher quality images.

Codes from Quadratic Residues

Quadratic residues can be used to construct a different code of length 12 that has more code words and corrects more errors than the repetition code described earlier. Table 7.4 lists the 24 code words in a code C_{12} of length 12. You can verify that any two code words in C_{12} disagree in at least six

Table 7.4: The 24 code words in the error-correcting code C_{12}

+ + − + + + − − − + − −	− − + − − − + + + − + +
− + + − + + + − − − + −	+ − − + − − − + + + − +
+ − + + − + + + − − − −	− + − − + − − − + + + +
− + − + + − + + + − − −	+ − + − − + − − − + + +
− − + − + + − + + + − −	+ + − + − − + − − − + +
− − − + − + + − + + + −	+ + + − + − − + − − − +
+ − − − + − + + − + + −	− + + + − + − − + − − +
+ + − − − + − + + − + −	− − + + + − + − − + − +
+ + + − − − + − + + − −	− − − + + + − + − − + +
− + + + − − − + − + + −	+ − − − + + + − + − − +
+ − + + + − − − + − + −	− + − − − + + + − + − +
− − − − − − − − − − − −	+ + + + + + + + + + + +

positions. A moment's thought reveals that this means C_{12} can correct up to two errors in each code word.

Example 3. The received transmission

differs from the first code word in Table 7.4 in the first two symbols and from every other code word in more than two symbols. Our best guess is that the first code word was the one actually transmitted.

Here is how the code C_{12} is constructed from the quadratic residues modulo 11. In the first code word,

$$+ + - + + + - - - + -,$$

the six "+" symbols occur in position 0 (the initial position) and in positions 1, 3, 4, 5, and 9—the quadratic residues modulo 11. Shift this 11-symbol pattern cyclically one position at a time to get the first 11 symbols of the next 10 code words. Then append the symbol "−" to each one. The last code word in the left column consists of 12 "−" symbols. Finally, the right column is obtained from the left by interchanging the symbols "+" and "−".

The same construction with the quadratic residues modulo the prime p produces the code C_{p+1} of length $p + 1$ and size $2(p + 1)$. It can be shown that the code corrects any configuration of at most $(p - 3)/4$ errors when $p + 1$ is a multiple of 4. The error-correcting codes actually used in modern space communication are more sophisticated than C_{p+1} but rely on similar principles.

7.4 Multiplication Tables, Legendre, and Euler

Quadratic residues are best understood through their in-

Table 7.5: Multiplication table modulo 13

×	1	2	3	4	5	6	7	8	9	10	11	12
1	1	2	3	4	5	6	7	8	9	10	11	12
2	2	4	6	8	10	12	1	3	5	7	9	11
3	3	6	9	12	2	5	8	11	1	4	7	10
4	4	8	12	3	7	11	2	6	10	1	5	9
5	5	10	2	7	12	4	9	1	6	11	3	8
6	6	12	5	11	4	10	3	9	2	8	1	7
7	7	1	8	2	9	3	10	4	11	5	12	6
8	8	3	11	6	1	9	4	12	7	2	10	5
9	9	5	1	10	6	2	11	7	3	12	8	4
10	10	7	4	1	11	8	5	2	12	9	6	3
11	11	9	7	5	3	1	12	10	8	6	4	2
12	12	11	10	9	8	7	6	5	4	3	2	1

teractions with one another, rather than in isolation. In the multiplication table modulo 13 (Table 7.5), each row and column contains the numbers 1, 2, ..., 12 in some order. (The table omits the uninteresting case of multiplication by 0.) The quadratic residues form a palindromic sequence along a diagonal.

The product uv of two quadratic residues u and v is also a quadratic residue since the two congruences

$$x^2 \equiv u \pmod{p} \qquad \text{and} \qquad y^2 \equiv v \pmod{p}$$

imply $(xy)^2 \equiv uv \pmod{p}$.

Table 7.6 reorders the rows and columns of the multiplication table modulo 13 to separate the quadratic residues

Table 7.6: Quadrants of quadratic residues and nonresidues

×		residues						nonresidues					
		1	3	4	9	10	12	2	5	6	7	8	11
residues	1	1	3	4	9	10	12	2	5	6	7	8	11
	3	3	9	12	1	4	10	6	2	5	8	11	7
	4	4	12	3	10	1	9	8	7	11	2	6	5
	9	9	1	10	3	12	4	5	6	2	11	7	8
	10	10	4	1	12	9	3	7	11	8	5	2	6
	12	12	10	9	4	3	1	11	8	7	6	5	2
nonresidues	2	2	6	8	5	7	11	4	10	12	1	3	9
	5	5	2	7	6	11	8	10	12	4	9	1	3
	6	6	5	11	2	8	7	12	4	10	3	9	1
	7	7	8	2	11	5	6	1	9	3	10	4	12
	8	8	11	6	7	2	5	3	1	9	4	12	10
	11	11	7	5	8	6	2	9	3	1	12	10	4

from the nonresidues. The residues occur in two 6-by-6 quadrants in the body of the table (shaded). The shaded upper left quadrant confirms that the product of two residues is a residue modulo 13. The other quadrants possess similar interpretations. These quadrants occur modulo any prime in accordance with the *multiplicative properties:*

- The product of two quadratic residues or of two quadratic nonresidues is a quadratic residue.
- The product of a quadratic residue and a nonresidue is a nonresidue.

The Legendre Symbol

In his pioneering investigations of quadratic residues, the French mathematician Adrien-Marie Legendre (1752–1833) introduced some inspired notation that greatly simplified his calculations. Multiplicativity tells us that products of quadratic residues and non-residues behave like products of positive and negative numbers (Table 7.7). To describe this behavior, Legendre mapped the quadratic residues and nonresidues to $+1$ and -1. More precisely, he defined what is now known as the *Legendre symbol* $(v\,|\,p)$ for each prime p as follows:

$$(v\,|\,p) = \begin{cases} +1 & \text{if } v \text{ is a quadratic residue modulo } p \\ -1 & \text{if } v \text{ is a quadratic nonresidue modulo } p \end{cases}$$

Table 7.7: Multiplication tables

×	residue	nonresidue
residue	residue	nonresidue
nonresidue	nonresidue	residue

×	+	−
+	+	−
−	−	+

Also, if v is divisible by p, then $(v\,|\,p) = 0$. We pronounce $(v\,|\,p)$ as "the Legendre symbol of v modulo p," or "the quadratic character of v modulo p."

Two familiar properties of the Legendre symbol take the following forms:

- multiplicativity: $(uv\,|\,p) = (u\,|\,p)(v\,|\,p)$.
- periodicity: If $v \equiv v' \pmod{p}$, then $(v\,|\,p) = (v'\,|\,p)$.

These properties often let us determine whether v is a quadratic residue modulo p with little effort.

Example 4. (a) The number 8 is a quadratic residue modulo 47 because

$$\begin{aligned}(8\,|\,47) &= (2\times 4\,|\,47) = (2\,|\,47)(4\,|\,47)\\ &= (49\,|\,47)(4\,|\,47) = 1\times 1 = 1.\end{aligned}$$

The computation relies on multiplicativity, periodicity, and the obvious fact that 4 and 49 are both quadratic residues modulo 47.

(b) A similar calculation shows that 27 is a quadratic residue modulo 47. Can you provide the details? Hint: $3 + 47 = 50$.

(c) Is 257 a quadratic residue modulo the prime 641? Because 257 is prime, it is difficult to exploit multiplicativity directly. We will revisit this example later.

(d) The two supplementary laws for quadratic reciprocity take the compact forms

$$(-1\,|\,p) = (-1)^{(p-1)/2} \qquad \text{and} \qquad (2\,|\,p) = (-1)^{(p^2-1)/8}.$$

Euler's Criterion

The Legendre symbol fits neatly with a test devised earlier by Leonhard Euler (1707–1783) to distinguish between quadratic residues and nonresidues.

Euler's criterion. If p is an odd prime, and ν is not divisible by p, then

$$\nu^{(p-1)/2} \equiv (\nu \mid p) \pmod{p}.$$

Example 5. (a) Table 7.8 confirms Euler's criterion for $p = 13$. It shows that $\nu^6 \equiv \pm 1 \pmod{13}$ for $\nu = 1, 2, \ldots, 12$ and that $\nu^6 \equiv 1 \pmod{13}$ exactly when ν is one of the quadratic residues 1, 3, 4, 9, 10, and 12.

Table 7.8: Euler's criterion modulo 13

ν	1	2	3	4	5	6	7	8	9	10	11	12
ν^6	1	−1	1	1	−1	−1	−1	−1	1	1	−1	1

(b) Is 10 a quadratic residue modulo 37? Euler's criterion tells us to look at 10^{18} modulo 37. The unwieldy computation can be shortened by noting that $10^3 \equiv 1000 \equiv 1 \pmod{37}$ since $999 = 37 \times 27$. Now

$$(10 \mid 37) \equiv 10^{18} \equiv \left(10^3\right)^6 \equiv 1^6 \equiv 1 \pmod{37}.$$

Therefore, 10 is a quadratic residue modulo 37.

(c) Is 257 a quadratic residue modulo the prime 641? Euler's criterion tells us to examine 257^{320} modulo 641. The computation is prohibitive in the absence of a shortcut of the type in (b). A better method will be given later.

(d) When $\nu = -1$, Euler's criterion asserts that

$$(-1 \mid p) \equiv (-1)^{(p-1)/2},$$

which is the first supplementary law again.

7.5 Some Square Roots

We have focused on squares modulo a prime. But what about *square roots*?

A Square Root Formula

To find the square roots of a given number s modulo p, we must assume that s is a quadratic residue modulo p. When $p \equiv -1 \pmod 4$, Euler's criterion and some algebra give

$$\left(\pm s^{(p+1)/4}\right)^2 \equiv s^{(p+1)/2} \equiv s \times s^{(p-1)/2} \equiv s \times (s|p) \pmod p.$$

We are assuming that $(s|p) = 1$, and so the two numbers $x = \pm s^{(p+1)/4}$ satisfy the congruence

$$x^2 \equiv s \pmod p.$$

We have found the square roots of s.

Square root formula. Suppose that the prime p satisfies $p \equiv -1 \pmod 4$. Then the square roots of the quadratic residue s modulo p are $\pm s^{(p+1)/4}$.

Example 6. (a) Let us find the square roots of 2 modulo the prime 103. Note that $103 \equiv -1 \pmod 4$. The square roots are $\pm 2^{26}$ modulo 103 by the square root formula. We can simplify this expression without dividing the huge number 2^{26} by 103 and taking the remainder. A shortcut lets us work with smaller numbers. The congruences

$$2^8 \equiv 256 \equiv 50 \pmod{103}$$

and $$2^9 \equiv 2 \times 2^8 \equiv 2 \times 50 \equiv 100 \equiv -3 \pmod{103}$$

give $2^{26} \equiv \left(2^9\right)^2 \times 2^8 \equiv (-3)^2 \times 50 \equiv 450 \equiv 38 \pmod{103}$. Therefore, the square roots of 2 modulo 103 are ± 38. It is easy to verify that $(\pm 38)^2 \equiv 1444 \equiv 2 \pmod{103}$.

(b) You can find the square roots of 2 modulo the prime 127 by a similar process using $2^7 = 128$.

By the way, when $p \equiv 1 \pmod 4$, the exponent $(p+1)/4$ is not an integer, and the square root formula does not work. No simple square root formula is known in this case.

More Square Roots

Each quadratic residue has exactly two square roots modulo a prime. However, the quadratic congruence

$$x^2 \equiv v \pmod m$$

can have more than two solutions when the modulus m is not a prime. For instance, 16 has *four* square roots modulo 209 since

$$(\pm 4)^2 \equiv 16 \pmod{209}$$

$$\text{and} \qquad (\pm 15)^2 \equiv 225 \equiv 16 \pmod{209}.$$

Note that $209 = 11 \times 19$, and the primes 11 and 19 are congruent to -1 modulo 4. Here is a general result along these lines.

Theorem. Let p and q be two primes congruent to -1 modulo 4 and let s be a square modulo pq that is divisible by neither p nor q. Then s has four square roots modulo pq.

The theorem guarantees the existence of four square roots but does not tell us how to find them. The Chinese Remainder Theorem (discussed in Section 6.5) does the job, as illustrated in the following example.

Example 7. Let us find the four square roots of $s = 8036$ modulo $m = 13081$. Note that $m = pq$, where the two primes $p = 103$ and $q = 127$ satisfy $p \equiv q \equiv -1 \pmod 4$.

Now $s = 8036 \equiv 2 \pmod{p}$, and the square roots of 2 modulo p are ± 38 by Example 6. Also, $s = 8036 \equiv 35 \pmod{q}$, and the square roots of 35 are ± 17 modulo q. So if x is a square root of s modulo m, then

$$x \equiv \pm 38 \pmod{p} \qquad \text{and} \qquad x \equiv \pm 17 \pmod{q}.$$

There are essentially two situations to treat. First, we seek a number x satisfying

$$x \equiv 38 \pmod{p} \qquad \text{and} \qquad x \equiv 17 \pmod{q}.$$

Example 3 in Chapter 6 solves these two simultaneous congruences with the Chinese Remainder Theorem. The solution is

$$x \equiv 11574 \pmod{m}.$$

We automatically get the solution $x \equiv -11574 \pmod{m}$ to the congruences

$$x \equiv -38 \pmod{p} \qquad \text{and} \qquad x \equiv -17 \pmod{q}.$$

Our first case has produced two of our four desired square roots modulo m, namely, ± 11574. These may also be written as ± 1507 modulo m.

For the second case, we seek a number x satisfying

$$x \equiv 38 \pmod{p} \qquad \text{and} \qquad x \equiv -17 \pmod{q}.$$

The Chinese Remainder Theorem gives

$$x \equiv 271 \pmod{m}.$$

Therefore, the four square roots of $s = 8036$ modulo $m = 103 \times 127$ are ± 1507 and ± 271.

The preceding extraction of the four square roots of s assumes we know the two prime factors p and q of m. Conversely, if we know the four square roots of s modulo m, then we can determine the two primes p and q. Problem 6 gives the details.

7.6 Marcia and Greg Flip a Coin

Sibling teenagers Marcia and Greg are arguing over who gets to use the family car one evening. They usually settle such disputes with a coin flip. Unfortunately, they have been banished to separate rooms and must communicate by an ordinary telephone. Neither of them trusts the other to report the result of a coin flip at the other end of the phone line. How can they conduct a coin flip fairly, with both parties satisfied that no one cheats? This is a problem in *cryptography,* the branch of mathematics that studies how to share and withhold information securely.

Marcia and Greg Do Number Theory

Fortunately, both Marcia and Greg know some number theory. Their knowledge of quadratic residues enables them to conduct a fair coin flip by telephone. Their conversation proceeds as follows.

Greg: Hi, Marcia. I have chosen two primes p and q, each of which is one less than a multiple of 4. Their product is $m = pq = 13081$. Marcia, if you can tell me p and q, then you win.

Marcia: Okay, Greg. I am choosing a number x with no factor in common with m. The square of x is $s = 8036$ modulo your number m. Greg, if you can guess x or $-x$ modulo m, then you win.

Greg: Let me see. I know that s is a quadratic residue modulo both p and q. (He expertly carries out the calculations in Example 7 and determines that the four square roots of s modulo m are ± 271 and ± 1507.) My guess is that the number you squared is $x = 1507$.

Marcia: Wrong! My number is $x = 271$. (She carries out her own computations here.) Your two primes are 103 and 127, Greg. The car is mine tonight!

Here is the mathematics behind the conversation. The number s has four square roots modulo m, only two of which are known to Marcia, namely, $\pm x$. She cannot readily find the other two square roots unless she factors m. But from his knowledge of p and q, Greg *can* determine that the four square roots of s are ± 217 and ± 1507 modulo m, as we did in Example 7. But he does not know whether Marcia selected $x = \pm 217$ or $x = \pm 1507$, and his guess is tantamount to calling heads or tails while a flipped coin is in the air. When he guesses incorrectly, Marcia can use the four square roots ± 217 and ± 1507 to figure out Greg's primes p and q. (See Problem 6 for details.) If Greg correctly guesses ± 217, then Marcia is unable to determine p and q, and she must concede.

Our discussion has been lighthearted, but the same issues arise in more serious cryptographic contexts. For instance, both the sender and the recipient of electronic certified mail with digital signatures need to confirm that the intended information is transmitted without tampering. Number theory provides the means to make such confirmations quickly and reliably.

Primes: Size Matters

The primes selected by Greg at the outset are unrealistically small. Marcia could factor m in a few moments by hand and instantly with a computer. To be safe, Greg should select p and q to be primes with many digits. For instance, if Greg informs Marcia that the product of his two primes p and q is

$$m = 85397342226737231891569795109543,$$

then she probably needs more information to determine p and q. She acquires that information only if Greg's guess for $\pm x$ is incorrect. Factoring large numbers into primes is a difficult problem—a key fact underlying our coin-flipping protocol. We point out that mere 50-digit numbers can now be factored relatively quickly with current technology. To be safe, Greg should select primes with hundreds of digits for p and q. The problem of finding primes of that magnitude is an interesting story in itself, which we omit.

7.7 Round Up at the Gauss Corral

A big breakthrough involving quadratic residues occurred in 1808 when the great German mathematician Carl Friedrich Gauss (1777–1855) transformed the computation of the Legendre symbol to a counting problem.

Gauss's lemma for quadratic residues. Let v be a positive integer that is not divisible by the odd prime p. Round each of the $(p-1)/2$ quotients

$$\frac{v}{p}, \frac{2v}{p}, \frac{3v}{p}, \ldots, \frac{\left(\frac{p-1}{2}\right)v}{p}$$

to the nearest integer. Let R be the number of quotients that round up. Then v is a quadratic residue modulo p if R is even, and a nonresidue if R is odd. In terms of the Legendre symbol,

$$(v \,|\, p) = (-1)^R.$$

Problem 18 shows how Gauss's lemma follows from Euler's criterion and explains why the mysterious *round-up counter R* appears.

The following example applies the lemma in some familiar situations.

Example 8. (a) When $v = 8$ and $p = 13$, the six quotients in Gauss's lemma are

$$\frac{8}{13}, \frac{16}{13}, \frac{24}{13}, \frac{32}{13}, \frac{40}{13}, \frac{48}{13}.$$

Three of these quotients, namely, 8/13, 24/13, and 48/13, round up to the nearest integer. Therefore, $R = 3$, and 8 is a quadratic nonresidue modulo 13, as asserted in Table 7.1.

(b) When $\nu = 1$, the quotients in Gauss's lemma are

$$\frac{1}{p}, \frac{2}{p}, \frac{3}{p}, \ldots, \frac{\frac{p-1}{2}}{p},$$

each of which rounds down to the nearest integer. Therefore, $R = 0$, and the lemma confirms the obvious fact that 1 is a quadratic residue modulo each prime p.

(c) When $\nu = p - 1$, the quotients in the lemma are

$$\frac{p-1}{p}, \frac{2(p-1)}{p}, \frac{3(p-1)}{p}, \ldots, \frac{\left(\frac{p-1}{2}\right)(p-1)}{p}.$$

Write

$$\frac{x(p-1)}{p} = x - \frac{x}{p},$$

and it becomes clear that all $(p-1)/2$ quotients round up to the nearest integer. So $R = (p-1)/2$ and

$$(-1 \mid p) = (p-1 \mid p) = (-1)^{(p-1)/2}.$$

We have found yet another confirmation of the first supplementary law.

Lattice Points in a Parallelogram

Gauss's lemma has a geometric interpretation that reveals a connection between quadratic residues and the selection of pixels representing a line on a computer monitor. The point (a, b) in the xy-plane is a *lattice point* provided the coordinates a and b are both integers. We will show that the round-up counter R in Gauss's lemma counts the lattice points in a parallelogram "corral" in the xy-plane. Equivalently, R counts certain dark pixels in the representation of a line segment on a computer monitor.

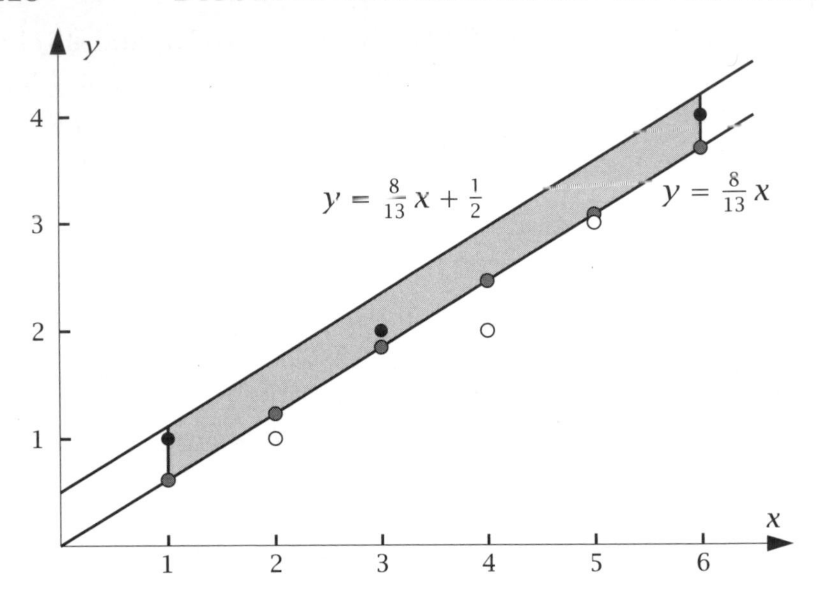

Figure 7.1: Gauss's quadratic residue lemma and lattice points

Here is a geometric interpretation of Example 8(a). When $v = 8$ and $p = 13$, the six quotients in Gauss's lemma are

$$\frac{8}{13}, \frac{16}{13}, \frac{24}{13}, \frac{32}{13}, \frac{40}{13}, \frac{48}{13}.$$

Figure 7.1 uses lattice points to depict the rounding process. The six points

$$\left(1, \tfrac{8}{13}\right), \left(2, \tfrac{16}{13}\right), \left(3, \tfrac{24}{13}\right), \left(4, \tfrac{32}{13}\right), \left(5, \tfrac{40}{13}\right), \left(6, \tfrac{48}{13}\right)$$

are shown as gray dots on the line $y = (8/13)\,x$. The vertical components of these points are our quotients, which are rounded down (white dots) or up (black dots) to the nearest lattice point. There are three black dots, and so $R = 3$. The three lattice points are rounded up and corralled in the shaded gray parallelogram bounded by the slanted lines

$$y = \left(\frac{8}{13}\right)x \qquad \text{and} \qquad y = \left(\frac{8}{13}\right)x + \frac{1}{2}$$

and the vertical lines $x = 1$ and $x = 6$.

The problem of rounding each of the six points

$$(x, y) = \left(x, \left(\tfrac{8}{13}\right) x\right) \qquad \text{for } x = 1, 2, \ldots, 6$$

to the nearest lattice point is the same problem solved by a computer in selecting the correct dark pixels to represent the line segment from $(0, 0)$ to $(13, 8)$. (Chapter 4 examines the computer line drawing problem in detail.)

Figure 7.2 superimposes the parallelogram in Figure 7.1 on the first seven dark pixels that represent the segment from $(0, 0)$ to $(13, 8)$. The lattice points are the centers of the pixels, and the three quotients that round up in Gauss's lemma correspond to the three pixel centers that fall above the segment.

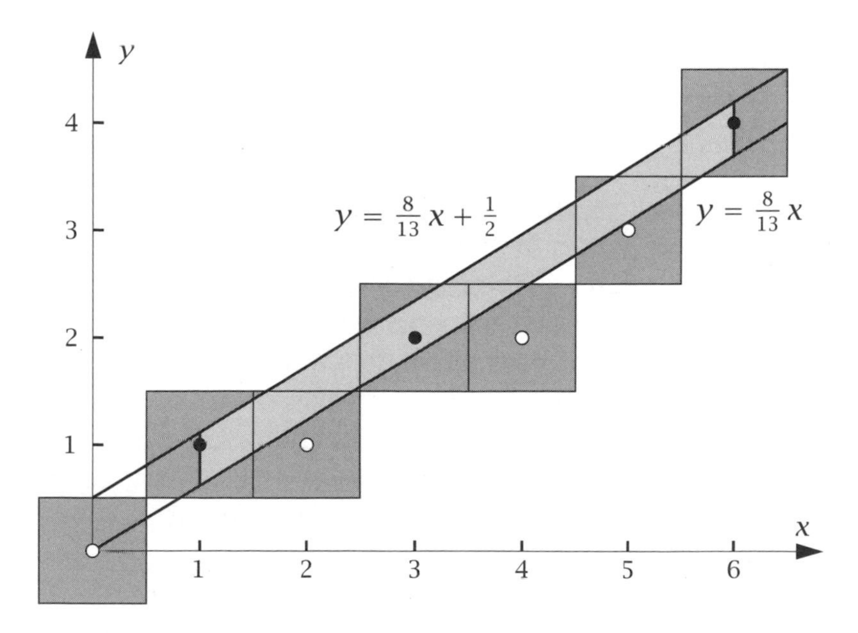

Figure 7.2: Gauss's quadratic residue lemma and line drawing

The general statement gives us two geometric interpretations for Gauss's round-up counter.

Theorem 1. Let p be an odd prime and let v be a positive integer not divisible by p.

(a) The round-up counter R in Gauss's lemma is equal to the number of lattice points in the parallelogram between the slanted lines

$$y = \left(\frac{v}{p}\right)x \quad \text{and} \quad y = \left(\frac{v}{p}\right)x + \frac{1}{2}$$

and the vertical lines $x = 1$ and $x = (p-1)/2$.

(b) Suppose that $1 \leq v \leq p - 1$. Represent the line segment joining $(0,0)$ to (p, v) with dark pixels on a computer monitor, as in Chapter 4. Then R is equal to the number of dark pixels (x, y) above the segment for $x = 1, 2, \ldots, (p-1)/2$.

Arithmetic Arrays Again

Chapter 4 outlines an elementary process involving arrays of numbers to determine the pixels a computer darkens to represent the segment joining the points $(0,0)$ and (v, p). Figure 7.3 shows the arithmetic array $A(8, -13)$ with the number $8x - 13y$ in position (x, y). In column x, the dark pixel is the one for which $8x - 13y$ has the smallest absolute value. The negative numbers in the array correspond to pixels above the line $y = (8/13)x$. It is now a simple matter to count the boxed entries and see that $R = 3$ when Theorem 1(b) is applied with $v = 8$ and $p = 13$.

In general, to represent the segment from $(0,0)$ to (v, p), we use the *arithmetic array* $A(v, -p)$ containing the number

$$vx - py$$

-104	-96	-88	-80	-72	-64	-56	-48	-40	-32	-24	-16	-8	0
-91	-83	-75	-67	-59	-51	-43	-35	-27	-19	-11	-3	5	13
-78	-70	-62	-44	-46	-38	-30	-22	-14	-6	2	10	18	26
-65	-57	-49	-41	-33	-25	-17	-9	-1	7	15	23	31	39
-52	-44	-36	-28	-20	-12	-4	4	12	20	28	36	44	52
-39	-31	-23	-15	-7	1	9	17	25	33	41	49	57	65
-26	-18	-10	-2	6	14	22	30	38	46	54	62	70	78
-13	-5	3	11	19	27	35	43	51	59	67	75	83	91
0	8	16	24	32	40	48	56	64	72	80	88	96	104

Figure 7.3: The arithmetic array $A(8, -13)$

in position (x, y). The dark pixel in column x is the one with smallest absolute value, and the round-up counter R equals the number of dark pixels with negative entries among columns $x = 1, 2, \ldots, (p-1)/2$.

Summary. The round-up counter R can be found by examining an arithmetic array and counting certain pixels.

Example 9. The arithmetic array $A(6, -7)$ is superimposed on the configuration of dark pixels representing the segment joining $(0, 0)$ and $(7, 6)$ in Figure 7.4. There are three dark pixels with negative entries for $x = 1, 2, 3$. Therefore, R is odd, and so 6 is a quadratic nonresidue modulo 7.

The situation depicted in Figure 7.4 holds for any odd prime p. The representation of the segment joining $(0, 0)$ and $(p, p-1)$ will have a staircase pattern of pixels. The

-42	-36	-30	-24	-18	-12	-6	0
-35	-29	-23	-17	-11	-5	1	7
-28	-22	-16	-10	-4	2	8	14
-21	-15	-9	-3	3	9	15	21
-14	-8	-2	4	10	16	22	28
-7	-1	5	11	17	23	29	35
0	6	12	18	24	30	36	42

Figure 7.4: The arithmetic array $A(6, -7)$

corresponding entries in the first $(p-1)/2$ columns of the arithmetic array $A(p-1, -p)$ are $-1, -2, \ldots, -(p-1)/2$. Thus $R = (p-1)/2$ in accordance with the first supplementary law. Problem 19 outlines a similar pixel-based verification of the second supplementary law.

7.8 It's the Law: Quadratic Reciprocity

Several eighteenth-century mathematicians noticed that if p and q are odd primes, then whether p is a quadratic residue modulo q depends in some manner on whether q is a quadratic residue modulo p. They worked empirically, organizing their data as in Table 7.9. We write "+" and "–" instead of $+1$ and -1 to make the patterns more apparent.

In most cases, $(p \mid q)$ and $(q \mid p)$ are equal, so that

$$(p \mid q)(q \mid p) = 1,$$

a relation reminiscent of the product of reciprocal fractions. The exceptions occur when p and q are both in the family

Table 7.9: The Legendre symbol $(p \mid q)$

$(p \mid q)$		q															
		3	5	7	11	13	17	19	23	29	31	37	41	43	47	53	59
p	3	0	−	−	+	+	−	−	+	−	−	+	−	−	+	−	+
	5	−	0	−	+	−	−	+	−	+	+	−	+	−	−	−	+
	7	+	−	0	−	−	−	+	−	+	+	+	−	−	+	+	+
	11	−	+	+	0	−	−	+	−	−	−	+	−	+	−	+	−
	13	+	−	−	−	0	+	−	+	+	−	−	−	+	−	+	−
	17	−	−	−	−	+	0	+	−	−	−	−	−	+	+	+	+
	19	+	+	−	−	−	+	0	−	−	+	−	−	−	−	−	+
	23	−	−	+	+	+	−	+	0	+	−	−	+	+	−	−	−
	29	−	+	+	−	+	−	−	+	0	−	−	−	−	−	+	+
	31	+	+	−	+	−	−	−	+	−	0	−	+	+	−	−	−
	37	+	−	+	+	−	−	−	−	−	−	0	+	−	+	+	−
	41	−	+	−	−	−	−	−	+	−	+	+	0	+	−	−	+
	43	−	−	−	+	+	+	−	+	−	+	−	+	0	+	+	+
	47	+	−	+	−	−	−	−	−	−	−	+	−	−	0	+	+
	53	−	−	+	+	+	+	−	−	+	−	+	−	+	+	0	+
	59	+	+	+	−	−	+	+	−	+	−	−	+	−	−	+	0

of primes 3, 7, 11, 19, 23, 31, 43, 47, 59, ... congruent to −1 modulo 4, in which case

$$(p \mid q)(q \mid p) = -1.$$

Table 7.10 re-organizes the data in Table 7.9 by separating the primes congruent to 1 modulo 4 from those congruent to −1 modulo 4. The resulting configuration gives our discovery a visually appealing form. The table is symmetric about the diagonal of 0's, except in the lower right quadrant, where the "+" and "−" symbols oppose one another. In other words, $(p \mid q) = (q \mid p)$ unless both p and q

Table 7.10: The Legendre symbol $(p \mid q)$ again

$(p \mid q)$		$q \equiv 1 \pmod 4$							$q \equiv -1 \pmod 4$								
		5	13	17	29	37	41	53	3	7	11	19	23	31	43	47	59
$p \equiv 1 \pmod 4$	5	0	−	−	+	−	+	−	−	−	+	+	−	+	−	−	+
	13	−	0	+	+	−	−	+	+	−	−	−	+	−	+	−	−
	17	−	+	0	−	−	−	+	−	−	−	+	−	−	+	+	+
	29	+	+	−	0	−	−	+	−	+	−	−	+	−	−	−	+
	37	−	−	−	−	0	+	+	+	+	+	−	−	−	−	+	−
	41	+	−	−	−	+	0	−	−	−	−	−	+	+	+	−	+
	53	−	+	+	+	+	−	0	−	+	+	−	−	−	+	+	+
$p \equiv -1 \pmod 4$	3	−	+	−	−	+	−	−	0	+	−	+	−	+	+	−	−
	7	−	−	−	+	+	−	+	−	0	+	−	+	−	+	−	−
	11	+	−	−	−	+	−	+	+	−	0	−	+	+	−	+	+
	19	+	−	+	−	−	−	−	−	+	+	0	+	−	+	+	−
	23	−	+	−	+	−	+	−	+	−	−	−	0	+	−	+	+
	31	+	−	−	−	−	+	−	−	+	−	+	−	0	−	+	+
	43	−	+	+	−	−	+	+	−	−	+	−	+	+	0	+	+
	47	−	−	+	−	+	−	+	+	+	−	−	−	−	−	0	+
	59	+	−	+	+	−	+	+	+	+	−	+	−	−	−	−	0

are congruent to -1 modulo 4. This property is the Law of Quadratic Reciprocity. Here is a formal statement.

Law of Quadratic Reciprocity. Let p and q be two odd primes. Then p is a quadratic residue modulo q if and only if q is a quadratic residue modulo p—unless both p and q are one less than a multiple of 4, in which case exactly one of p and q is a quadratic residue of the other. In terms of the Legendre symbol,

$$(q \mid p)(p \mid q) = (-1)^{(p-1)(q-1)/4}.$$

Note that the exponent $(p-1)(q-1)/4$ is even unless both p and q are congruent to -1 modulo 4.

Applications of Quadratic Reciprocity

Some examples demonstrate the power of quadratic reciprocity.

Example 10. (a) The following computations show that the prime 257 is a quadratic nonresidue modulo the prime 641, a fact that stymied us earlier:

$$\begin{aligned}(257\,|\,641) &= (641\,|\,257) = (127\,|\,257) = (257\,|\,127)\\ &= (3\,|\,127) = -(127\,|\,3) = -(1\,|\,3) = -1.\end{aligned}$$

We used periodicity to replace $(641\,|\,257)$ with $(127\,|\,257)$ and to replace $(257\,|\,127)$ with $(3\,|\,127)$. We used reciprocity several times to replace $(p\,|\,q)$ with $\pm(q\,|\,p)$. The negative sign arises because the primes 3 and 127 are both congruent to -1 modulo 4.

(b) Is 282 a quadratic residue modulo the prime 1597? Start with the factorization $282 = 2 \times 3 \times 47$ and write

$$\begin{aligned}(282\,|\,1597) &= (2 \times 3 \times 47\,|\,1597)\\ &= (2\,|\,1597)(3\,|\,1597)(47\,|\,1597).\end{aligned}$$

Now analyze each factor in turn. The second supplementary law gives

$$(2\,|\,1597) = -1.$$

since $1597 \equiv 5 \pmod 8$. By quadratic reciprocity and periodicity,

$$(3\,|\,1597) = (1597\,|\,3) = (1\,|\,3) = 1.$$

Finally, quadratic reciprocity, periodicity, and the first supplementary law give

$$(47\,|\,1597) = (1597\,|\,47) = (46\,|\,47) = (-1)^{(47-1)/2} = -1.$$

Combine all the factors to see that

$$(282 \mid 1597) = (-1) \times 1 \times (-1) = 1.$$

Therefore, 282 is a quadratic residue modulo 1597.

Quadratic reciprocity helps identify the primes for which a given number v is a quadratic residue. In other words, we now have a way to attack the second fundamental question about quadratic residues. Let us illustrate the method for $v = 3$.

Example 11. When is 3 a quadratic residue modulo the odd prime p? In other words, for which p does $(3|p)$ equal 1? Reciprocity gives

$$(3|p)(p|3) = (-1)^{(p-1)(3-1)/4} = (-1)^{(p-1)/2}.$$

It follows that $(3|p) = 1$ if and only if $(p|3)$ and $(-1)^{(p-1)/2}$ are both 1 or both -1. In the former case, we must have

$$p \equiv 1 \pmod 3 \qquad \text{and} \qquad p \equiv 1 \pmod 4.$$

The two congruences tell us that $p \equiv 1 \pmod{12}$. In the latter case,

$$p \equiv -1 \pmod 3 \qquad \text{and} \qquad p \equiv -1 \pmod 4,$$

and so $p \equiv -1 \pmod{12}$. Therefore, 3 is a quadratic residue modulo p if and only if p is one more or one less than a multiple of 12.

A Proof of Quadratic Reciprocity: Count the Dots

The phenomenon of quadratic reciprocity was discovered empirically by both Euler and Legendre. However, neither of them was able to supply a rigorous proof that reciprocity held for all pairs of primes. Finally, Gauss proved the Law of Quadratic Reciprocity in 1796 at age 19. His first proof was long and somewhat awkward. Later in his career, he returned to give several elegant demonstrations of this *theorema aureum* (golden theorem), as he called it.

Figure 7.5 illustrates one proof of quadratic reciprocity that relies on Gauss's lemma and a few ideas of Ferdinand Eisenstein (1823–1852), one of Gauss's students. In the figure $p = 19$ and $q = 11$, but the same method works generally. By Gauss's lemma

$$(q\,|\,p)(p\,|\,q) = (-1)^R(-1)^S = (-1)^{R+S},$$

where R is the number of lattice points in the parallelogram

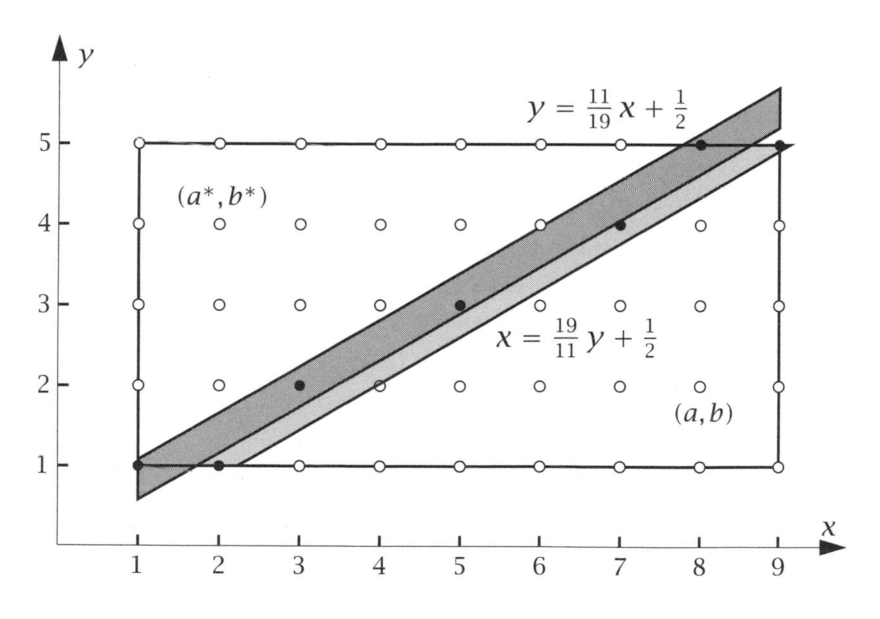

Figure 7.5: Lattice points and quadratic reciprocity for $p = 19$ and $q = 11$. The line $y = (11/19)x$ separates the two shaded parallelograms.

defined by

$$\left(\frac{q}{p}\right)x \le y \le \left(\frac{q}{p}\right)x + \frac{1}{2} \qquad \text{and} \qquad 1 \le x \le \frac{p-1}{2},$$

and S is the number of lattice points in the parallelogram defined by

$$\left(\frac{p}{q}\right)x \le y \le \left(\frac{p}{q}\right)x + \frac{1}{2} \qquad \text{and} \qquad 1 \le x \le \frac{q-1}{2}.$$

The first parallelogram is shown in dark gray in the figure. We reflect the second parallelogram about the line $y = x$ by interchanging x and y in the defining inequalities to get

$$\left(\frac{p}{q}\right)y \le x \le \left(\frac{p}{q}\right)y + \frac{1}{2} \qquad \text{and} \qquad 1 \le y \le \frac{q-1}{2}.$$

This reflected parallelogram (in light gray) also contains S lattice points and fits snugly under the first parallelogram. In the figure $R = 4$ and $S = 3$.

The lattice points of both parallelograms in the figure occur in a rectangle with $(p-1)/2$ lattice points on a horizontal edge and $(q-1)/2$ lattice points on a vertical edge. We focus on the lattice points of the rectangle that fall outside the parallelograms in the two white regions. Let N denote the number of these lattice points in the lower right white region. In the figure $N = 19$. Each lattice point (a, b) in the lower right region is paired with a symmetric mate (a^*, b^*) in the upper left region, where

$$a + a^* = \frac{p+1}{2} \qquad \text{and} \qquad b + b^* = \frac{q+1}{2}.$$

So there are $2N$ lattice points outside the two parallelograms in the rectangle. We now make three straightforward observations.

- The numbers $R + S$ and $R + S + 2N$ are both even or both odd.
- The rectangle has a total of $R + S + 2N$ lattice points.
- The rectangle has a total of

$$\frac{p-1}{2} \times \frac{q-1}{2} = \frac{(p-1)(q-1)}{4}$$

lattice points.

Put all the pieces together to see that

$$(q \mid p)(p \mid q) = (-1)^{R+S} = (-1)^{R+S+2N} = (-1)^{(p-1)(q-1)/4},$$

which is the Law of Quadratic Reciprocity.

7.9 Notes and References

Books that discuss primes and applications include [2, 3, 5]. Quadratic residues and allied topics are covered in many number theory books, including the classic one by Davenport [4]. Stillwell's excellent books [11] and [12] treat most of the topics in this chapter and give a different proof of quadratic reciprocity.

The Law of Quadratic Reciprocity lies at the crossroads of many areas of advanced mathematics, and scores of diverse proofs have been discovered based on techniques from those fields. The site

`www.rzuser.uni-heidelberg.de/~hb3/fchrono.html`

lists and classifies the known proofs. Eisenstein's approach to the Law of Quadratic Reciprocity goes beyond what we presented and is fully displayed in a clear and entertaining manner in [6].

The error-correcting codes we constructed with quadratic residues are by no means the best codes known. See [7] for the construction of many other codes. Our discussion of coin flips and quadratic residues is based on [1, 9, 10] and Chapter 2 of [8].

1. Blum, L., Blum, M., and Shub, M., A simple unpredictable pseudo-random number generator, *SIAM Journal on Computing* **15** (1986), 364-383.
2. Burn, R. P., *A Pathway into Number Theory.* Cambridge University Press, Cambridge, UK, 1982.

3. Crandall, Richard, and Carl Pomerance, Carl, *Prime Numbers,* 2nd ed. Springer, New York, 2005.
4. Davenport, H., *The Higher Arithmetic,* 8th ed. Cambridge University Press, Cambridge, UK, 2008.
5. Hankerson, D. R., Hoffman, D. G., Leonard, D. A., Lindner, C. C., Phelps, K. T., Rodger, C. A., and Wall, J. R., *Coding Theory and Cryptography,* 2nd ed. Marcel Dekker, New York, 2000.
6. Laubenbacher, R. C., and Pengelley, D. J., *Mathematical Intelligencer* **16** (1994), 67–72.
7. Moon, Todd K., *Error Correction Coding: Mathematical Methods and Algorithms.* Wiley-Interscience, Hoboken, New Jersey, 2005.
8. Peterson, Ivars, *The Mathematical Tourist.* W. H. Freeman, New York, 1988.
9. Rabin, M. O., Digitalized signatures and public-key functions as intractable as factorization, Technical Report: TR-212, 1979, Massachusetts Institute of Technology, Cambridge, Massachusetts.
10. Shepherd, S. J., Sanders, P. W., and Stockel, C. T., The quadratic residue cipher and some notes on implementation, *Cryptologia* **17** (1993), 264-282.
11. Stillwell, John, *Elements of Number Theory.* Springer, New York, 2003.
12. Stillwell, John, *Numbers and Geometry.* Springer, New York, 1998.

7.10 Problems

1. (a) Let the prime p be one more than a multiple of 4. Explain why v and $-v$ are both quadratic residues or quadratic nonresidues modulo p.

   (b) What happens if p is one less than a multiple of 4?

2. The *Leyland number* $p = 4405^{2638} + 2638^{4405}$ was shown to be prime in 2004.

   (a) Is $p - 1$ a quadratic residue modulo p?

   (b) Is 2 a quadratic residue modulo p?

3. Consider the 24 code words of length 12 in the code C_{12} we constructed from the quadratic residues modulo 11. See Table 7.4.

(a) If the received string is

$$- - - - - + + + - + -,$$

what word was most likely sent?

(b) Show that if at most two errors occur during transmission, then there are

$$24(1 + 12 + 66) = 1896$$

strings of 12 symbols that could be received. Hint: How is the choose function $\binom{n}{k}$ defined? (See Chapter 1.)

4. (a) Find the two numbers x modulo 103 satisfying the quadratic congruence

$$x^2 + 6x + 7 \equiv 0 \pmod{103}.$$

Hint: Complete the square. Use Example 6.

(b) Explain why no number x satisfies the quadratic congruence

$$x^2 + 6x + 10 \equiv 0 \pmod{103}.$$

(c) Find a value of a so that the congruence

$$x^2 + 6x + a \equiv 0 \pmod{103}$$

has exactly one solution x modulo 103.

5. Explain how Marcia can check that her choice of x has no factor in common with Greg's number m, even though she does not know how to factor m into primes in Section 7.6.

6. This problem outlines how Marcia determines Greg's primes p and q from her knowledge of the square roots $x = \pm 271$ and $y = \pm 1507$ of $s = 8036$ modulo $m = pq = 13081$ in Section 7.6. We know that $x^2 \equiv y^2 \equiv s \pmod{m}$.

(a) Show that $(x + y)(x - y) \equiv 0 \pmod{m}$.

(b) Explain why $x + y$ is divisible by exactly one of the two primes p and q.

(c) Use Euclid's algorithm to compute the greatest common divisor of $x + y$ and m.

(d) Explain how the above calculations give Marcia one of Greg's primes.

(e) Explain how Marcia can find Greg's other prime.

7. Show that 1729 is a quadratic nonresidue modulo the prime 8923. Hint: $1729 = 7 \times 13 \times 19$.

8. Show that 641 is a quadratic nonresidue modulo *Jenny's prime* 8675309 (named for a woman's phone number in a catchy song of the 1980s by Tommy Tutone).

9. Show that 5 is a quadratic residue modulo the odd prime p if and only if $p \equiv \pm 1 \pmod{10}$.

10. What can you say about the odd prime p if 7 is a quadratic residue modulo p?

11. What can you say about the odd prime p if 11 is a quadratic residue modulo p?

12. Show that -2 is a quadratic residue modulo the odd prime p if and only if $p \equiv 1$ or $3 \pmod{8}$.

13. What can you say about the odd prime p if -3 is a quadratic residue modulo p? Hint: One approach relies on multiplicativity and Example 11.

14. What can you say about the odd prime p if -5 is a quadratic residue modulo p?

15. Explain how Euler's criterion implies Fermat's congruence: $v^p \equiv v \pmod{p}$ for each integer v and each prime p. See Section 5.8.

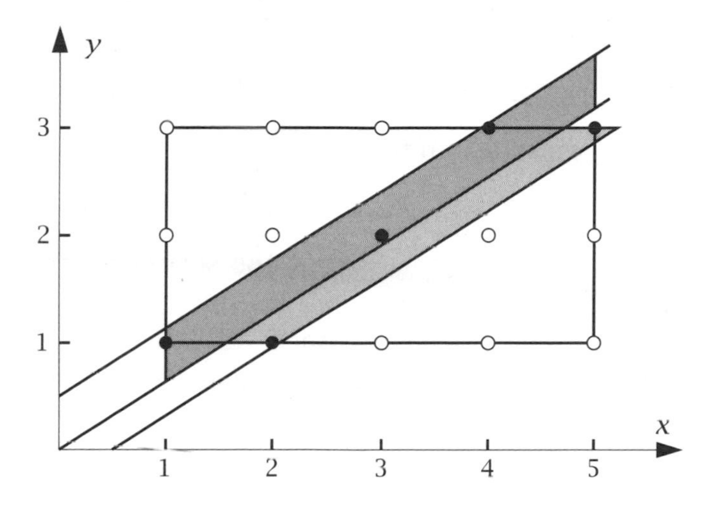

Figure 7.6: Quadratic reciprocity for two primes p and q

16. Figure 7.6 illustrates our proof of the Law of Quadratic Reciprocity for two primes p and q. What are p and q? Find the values of R, S, and N.

17. Let $\overline{a}$ be the remainder when the positive integer a is divided by a prime p. Thus $0 \le \overline{a} \le p-1$. Explain why the round-up counter R in Gauss's lemma is equal to the number of remainders in the list

$$\overline{v},\ \overline{2v},\ \overline{3v},\ \ldots,\ \overline{\left(\tfrac{p-1}{2}\right) v}$$

that are greater than $(p-1)/2$.

18. This problem outlines a proof of Gauss's lemma for quadratic residues. Let v be a positive integer not divisible by the odd prime p. Let R be the corresponding round-up counter.

    (a) Show that no two of the $(p-1)/2$ factors in the following product are congruent modulo p or sum to 0 modulo p:

$$Q = v \times 2v \times 3v \times \cdots \times \left(\tfrac{p-1}{2}\right) v.$$

(b) Use Problem 17 to explain why

$$Q \equiv 1 \times 2 \times 3 \times \cdots \times \left(\tfrac{p-1}{2}\right)(-1)^R \pmod{p}.$$

(c) Apply Euler's criterion to deduce Gauss's lemma:

$$(\nu \mid p) \equiv \nu^{(p-1)/2} \equiv (-1)^R \pmod{p}.$$

19. Give a pixel-based confirmation of the second supplementary law similar to the one used for the first supplementary law in Example 9. The arithmetic arrays $A(2,-11)$ and $A(2,-13)$ in Figure 7.7 illustrate the cases $p \equiv \pm 3 \pmod{8}$.

-22	-20	-18	-16	-14	-12	-10	-8	-6	-4	-2	0
-11	-9	-7	-5	-3	-1	1	3	5	7	9	11
0	2	4	6	8	10	12	14	16	18	20	22

-26	-24	-22	-20	-18	-16	-14	-12	-10	-8	-6	-4	-2	0
-13	-11	-9	-7	-5	-3	-1	1	3	5	7	9	11	13
0	2	4	6	8	10	12	14	16	18	20	22	24	26

Figure 7.7: The arithmetic arrays $A(2,-11)$ and $A(2,-13)$

If error is corrected whenever it is recognized as such,
the path of error is the path to truth.
HANS REICHENBACH

It is not knowledge, but the act of learning,
not possession but the act of getting there,
which grants the greatest enjoyment.
C. F. GAUSS

References

Alexanderson, G. L., and Wetzel, J. E., Simple partitions of space, *Mathematics Magazine* **51** (1978), 220–225.

Bailey, H., Fetching water with least residues, *College Mathematics Journal* **39** (2008), 304–306.

Ball, Keith, *Strange Curves, Counting Rabbits, and Other Mathematical Explorations.* Princeton University Press, Princeton, New Jersey, 2003.

Banks, Robert B., *Slicing Pizzas, Racing Turtles, and Further Adventures in Applied Mathematics.* Princeton University Press, Princeton, New Jersey, 1999.

Beck, Anatole, Bleicher, Michael, N., and Crowe, Donald W., *Excursions into Mathematics.* A. K. Peters, Natick, Massachusetts, 2000.

Beck, Matthias, and Robins, Sinai, *Computing the Continuous Discretely: Integer-Point Enumeration in Polyhedra.* Springer, New York, 2006.

Beihoffer, D., Hendry, J., Nijenhuis, A., and Wagon, S., Faster algorithms for Frobenius numbers, *Electronic Journal of Combinatorics* **12** (2005), #R27 38 pp (electronic).

Bellman, R., Cooke, K. L., and Lockett, J. A., *Algorithms, Graphs, and Computers.* Academic Press, New York, 1970.

Berlekamp, Elwyn, R., Conway, John H., and Guy, Richard K., *Winning Ways for Your Mathematical Plays.* Academic Press, London, 1985.

Beutelspacher, Albert, *Cryptology.* Mathematical Association of America, Washington, DC, 1994.

Blatter, C., Another proof of Pick's area theorem, *Mathematics Magazine* **70** (1974), 200.

Blum, L., Blum, M., and Shub, M., A simple unpredictable pseudorandom number generator, *SIAM Journal on Computing* **15** (1986), 364–383.

Boldi, P., Santini, M., Vigna, S., Measuring with jugs, *Theoretical Computer Science* **282** (2002), 259–270.

Brauer, A., On a problem of partitions, *American Journal of Mathematics* **64** (1942), 299–312.

Bresenham, J. E., Algorithm for computer control of a digital plotter, *IBM Systems Journal* **4** no. 1 (1965), 25–30. Also available at: `www.research.ibm.com/journal/sj/041/ibmsjIVRIC.pdf`

Brualdi, Richard A., *Introductory Combinatorics,* 4th ed. Pearson/Prentice-Hall, Upper Saddle River, New Jersey, 2004.

Bruckenheimer M. and Arcavi, A., Farey series and Pick's area theorem, *Mathematical Intelligencer* **17** (1995), 200.

Burger, Edward B., and Starbird, Michael, *The Heart of Mathematics.* Key College Publishing, Emeryville, California, 2000.

Burn, R. P., *A Pathway into Number Theory.* Cambridge University Press, Cambridge, U,K., 1982.

Chartrand, Gary, and Lesniak, Linda, *Graphs & Digraphs,* 4th ed. Chapman & Hall/CRC, Boca Raton, Florida, 2005.

Chvátal, V., A combinatorial theorem in plane geometry, *Journal of Combinatorial Theory, Series B* **18** (1975), 39–41.

Coxeter, H. S. M., and Greitzer, S. L., *Geometry Revisited.* Mathematical Association of America, Washington, DC, 1967.

Crandall, Richard, and Pomerance, Carl, *Prime Numbers,* 2nd ed. Springer, New York, 2005.

Cromwell, Peter R., *Polyhedra,* New ed. Cambridge University Press, Cambridge, U.K., 1999.

Cuoco, Al, *Mathematical Connections.* Mathematical Association of America, Washington, DC, 2005.

Czyzowicz, J., Rivera-Campo, E., Santoro, N., Urrutia, J., and Zaks, J., Tight bounds for the rectangular art gallery problem, *Graph-Theoretic Concepts in Computer Science (Fischbachau 1991),* 105–112, Lecture Notes in Computer Science 570, Springer, Berlin, 1992.

Davenport, H., *The Higher Arithmetic,* 7th ed. Cambridge University Press, Cambridge, U.K., 2000.

DeTemple D., and Robertson, J. M., The equivalence of Euler's and Pick's theorems, *Mathematics Teacher* **67** (1974), 222–226.

Ehrhart, E., Sur les polyèdres rationnels homothétiques à *n* dimensions, *Comptes Rendus de l'Academie des Science* **254** (1962), 616–618.

Ehrhart, E., *Polynómes arithmétiques et Méthode des Polyèdres en Combinatoire.* International Series of Numerical Mathematics, vol. 35, Birkhäuser Verlag, Basel/Stuttgart, 1977.

Fisher, D. C., Collins, K. L., and Krompart, L. B., Problem 10406 and solution, *American Mathematical Monthly* **104** (1997), 572–573.

Fisk, S., A short proof of Chvátal's watchman theorem, *Journal of Combinatorial Theory, Series B* **24** (1978), 374.

Funkenbusch, W. W., From Euler's formula to Pick's formula using an edge theorem, *American Mathematical Monthly* **81** (1974), 647–648.

Gaskell, R. W., Klamkin, M. S., and Watson, P., Triangulations and Pick's theorem, *Mathematics Magazine* **49** (1976), 35–37.

Glassner, Andrew S., (ed.), *Graphics Gems.* Academic Press Professional, Boston, 1990.

Grimaldi, Ralph P., *Discrete and Combinatorial Mathematics: An Applied Introduction,* 5th ed. Addison Wesley Longman, Reading, Massachusetts, 2004.

Grünbaum B., and Shephard, G. C., Pick's theorem, *American Mathematical Monthly* **100** (1993), 150–161.

Hankerson, D. R. Hoffman, D. G., Leonard, D. A., Lindner, C. C., Phelps, K. T., Rodger, C. A., and Wall, J. R., *Coding Theory and Cryptography,* 2nd ed. Marcel Dekker, New York, 2000.

Harris M. A., and Reingold, E. M., Line drawing, leap years, and Euclid, *ACM Computing Surveys* **36** (2004), 68–80.

Hess, R. I., Bonus problem, *The Bent of Tau Beta Pi* **96** (2005), No. 3, 53.

Honsberger, Ross, *Mathematical Gems II.* Mathematical Association of America, Washington, DC, 1976.

Iseri, H., An exploration of Pick's theorem in space, *Mathematics Magazine* **81** (2008), 106–115.

Ismailescu, D., Slicing the pie, *Discrete and Computational Geometry* **30** (2003), 263–276.

Kahn, J., Klawe, M., and Kleitman, D., Traditional galleries require fewer watchmen, *SIAM Journal of Algebraic and Discrete Methods* **4** (1983), 194–206.

Kertzner, S., The linear diophantine equation, *American Mathematical Monthly* **88** (1981), 200–203.

Larsen, L., A discrete look at $1 + 2 + \cdots + n$, *College Mathematics Journal* **16** (1985), 369–382.

Laubenbacher R. C. and Pengelley, D. J., *Mathematical Intelligencer* **16** (1994), 67–72.

Liu, A. C. F., Lattice points and Pick's theorem, *Mathematics Magazine* **52** (1979), 232–235.

Lovász, László, Pelikán, József, and Vesztergombi, Katalin, *Discrete Mathematics: Elementary and Beyond.* Springer, New York, 2003.

Macdonald, I. G., The volume of a lattice polyhedron, *Proceedings of the Cambridge Philosophical Society* **59** (1963), 719–726.

McDiarmid, C. J. H., and Ramírez Alfonsín, J. L., Sharing jugs of wine, *Discrete Mathematics* **125** (1994), 279–287.

Michael, T. S., and Pinciu, V., Art gallery theorems for guarded guards, *Computational Geometry* **26** (2003), 247–258.

Moon, Todd K., *Error Correction Coding: Mathematical Methods and Algorithms.* Wiley-Interscience, Hoboken, New Jersey, 2005.

Nelson, Roger B., *Proofs without Words: Exercises in Visual Thinking.* Mathematical Association of America, Washington, DC, 1993.

Nelson, Roger B., *Proofs without Words II: More Exercises in Visual Thinking.* Mathematical Association of America, Washington, DC, 2000.

Nijenhuis, A., and Wilf, H. S., Representation of integers by linear forms in non-negative integers, *Journal of Number Theory* **4** (1972), 98–106.

Niven, Ivan, *Diophantine Approximations.* Dover, New York, 2008.

Nyblom, M. A., Pick's theorem and greatest common divisors, *Mathematical Spectrum* **38** (2005/06), 9–11.

O'Bierne, T. H., *Puzzles and Paradoxes.* Oxford University Press, New York, 1965.

Ogilvy, C. Stanley, and Anderson, John T., *Excursions in Number Theory.* Dover, New York, 1988.

O'Rourke, J., Galleries need fewer mobile guards: a variation on Chvátal's theorem, *Geometriae Dedicata* **14** (1983), 273–283.

O'Rourke, Joseph, *Art Gallery Theorems.* Oxford University Press, Cambridge, U.K., 1987.

O'Rourke, Joseph, *Computational Geometry in C,* 2nd ed. Cambridge University Press, Cambridge, U.K., 1998.

Owens, R. W., An algorithm to solve the Frobenius problem, *Mathematics Magazine* **76** (2003), 264–275.

Parshall, Karen Hunger, *James Joseph Sylvester: Jewish Mathematician in a Victorian World.* Johns Hopkins University Press, Baltimore, 2006.

Peterson, Ivars, *The Mathematical Tourist.* W. H. Freeman, New York, 1988.

Pfaff, T. J., and Tran, M. M., The generalized jug problem, *Journal of Recreational Mathematics* **31** (2002–03), 100–103.

Pick, G., Geometrisches zur Zahlenlehre, *Sitzungsberichte Lotos (Prag) Naturwissenschaftlich-Medizinschen Vereines für Böhmen* **19** (1899), 311–319.

Rabin, M. O., Digitalized signatures and public-key functions as intractable as factorization, Technical Report: TR-212, 1979, Massachusetts Institute of Technology, Cambridge, Massachusetts.

Ramírez Alfonsín, J. L., *The Diophantine Frobenius Problem.* Oxford University Press, New York, 2005.

Reeve, J. E., On the volume of lattice polyhedra, *Proceedings of the London Mathematical Society* (3rd Ser.), **7** (1957), 378–395.

Reingold, Edward M., and Dershowitz, Nachum, *Calendrical Calculations: The Millennium Edition.* Cambridge University Press, Cambridge, U.K., 2001.

Reingold, Edward M., and Dershowitz, Nachum, *Calendrical Tabulations, 1900–2200.* Cambridge University Press, Cambridge, U.K., 2002.

Richeson, David S., *Euler's Gem: The Polyhedron Formula and the Birth of Topology.* Princeton University Press, Princeton, New Jersey, 2008.

Roberts, Fred S., and Tesman, Barry, *Applied Combinatorics,* 2nd ed. Prentice Hall, Upper Saddle River, New Jersey, 2003.

Roberts, S., On the figures formed by the intercepts of a system of straight lines in a plane, and on analogous relations in space of three dimensions, *Proceedings of the London Mathematical Society* **19** (1899), 405–422.

Rosenholtz, I., Calculating surface areas from a blueprint, *Mathematics Magazine* **52** (1979), 252–256.

Sack, J.-R., and Urrutia, J. (eds.), *Handbook of Computational Geometry.* North Holland, Amsterdam, 2000.

Sawyer, W. W., On a well-known puzzle, *Scripta Mathematica* **16** (1950), 107–110.

Schumer, Peter D., *Mathematical Journeys.* Wiley Interscience, Hoboken, New Jersey, 2004.

Shepherd, S. J., Sanders, P. W., and Stockel, C. T., The quadratic residue cipher and some notes on implementation, *Cryptologia* **17** (1993), 264–282.

Sicherman, G., Theory and practice of Sylver Coinage, *Integers: Electronic Journal of Combinatorial Number Theory* **2** (2002), #G02 11 pp (electronic).

Sloane, N. J. A., The On-Line Encyclopedia of Integer Sequences, *Notices of the American Mathematical Society* **50** (2003), 912–915.

Steiner, J., Eineige Gesetze über die Theilung der Ebene und des Raumes, *J. für die Reine und Angewandte Mathematik* **1** (1826), 349–364.

Stewart, Ian, *Another Fine Math You've Got Me Into. . . .* Dover, New York, 2003.

Stillwell, John, *Elements of Number Theory.* Springer, New York, 2003.

Stillwell, John, *Numbers and Geometry.* Springer, New York, 1998.

Stinson, Douglas, *Cryptology: Theory and Practice,* 2nd ed. Chapman Hall/CRC, Boca Raton, Florida, 2002.

Sylvester, J. J., On subinvariants, i.e. semi-invariants to binary quantics of an unlimited order, *American Journal of Mathematics* **5** (1882), 119–136.

Tweedie, M. C. K., A graphical method of solving Tartaglian measuring puzzles, *Mathematical Gazette* **23** (1939), 278–282.

Varberg, D. E., Pick's theorem revisited, *American Mathematical Monthly* **92** (1985), 584–587.

Weaver, C. S., Geoboard triangles with one interior point, *Mathematics Magazine* **50** (1977) 92–94.

Wetzel, J. E., On the division of the plane by lines, *American Mathematical Monthly* **85** (1978), 647–656.

Wilf, H. S., A circle-of-lights algorithm for the "money-changing problem," *American Mathematical Monthly* **85** (1978), 562–565.

Wu, X., and Rokne, J. G., Double-step incremental generation of lines and circles, *Computer Vision, Graphics, and Image Processing* **37** (1987), 331–334.

Zimmerman, S., Slicing space, *College Mathematics Journal* **32** (2001), 126–128.

Żyliński, P., Placing guards in art galleries by graph coloring, *Contemporary Mathematics* **352** (2004), 177–188.

Index